*Proceedings of the Thirty-First
Annual Biology Colloquium*

Dr. Wiens   Dr. Odum   Dr. McIntire

Dr. Likens

Dr. Odum

Dr. Pielou

Dr. Riley   Dr. Golley   Dr. Whittaker

Ecosystem Structure and Function

The Annual Biology Colloquium

YEAR, THEME, AND LEADER

1939. *Recent Advances in Biological Science.* Charles Atwood Kofoid
1940. *Ecology.* Homer LeRoy Shantz
1941. *Growth and Metabolism.* Cornelis Bernardus van Niel
1942. *The Biologist in a World at War.* William Brodbeck Herns
1943. *Contributions of Biological Science to Victory.* August Leroy Strand
1944. *Genetics and the Integration of Biological Sciences.* George Wells Beadle
1945. (Colloquium canceled)
1946. *Aquatic Biology.* Robert C. Miller
1947. *Biogeography.* Ernst Antevs
1948. *Nutrition.* Robert R. Williams
1949. *Radioisotopes in Biology.* Eugene M. K. Geiling
1950. *Viruses.* W. M. Stanley
1951. *Effects of Atomic Radiations.* Curt Stern
1952. *Conservation.* Stanley A. Cain
1953. *Antibiotics.* Wayne W. Umbreit
1954. *Cellular Biology.* Daniel Mazia
1955. *Biological Systematics.* Ernst Mayr
1956. *Proteins.* Henry Borsook
1957. *Arctic Biology.* Ira Loren Wiggins
1958. *Photobiology.* F. W. Went
1959. *Marine Biology.* Dixy Lee Ray
1960. *Microbial Genetics.* Aaron Novick
1961. *Physiology of Reproduction.* Frederick L. Hisaw
1962. *Insect Physiology.* Dietrich Bodenstein
1963. *Space Biology.* Allan H. Brown
1964. *Microbiology and Soil Fertility.* O. N. Allen
1965. *Host-Parasite Relationships.* Justus F. Mueller
1966. *Animal Orientation and Navigation.* Arthur Hasler
1967. *Biometeorology.* David M. Gates
1968. *Biochemical Coevolution.* Paul R. Ehrlich
1969. *Biological Ultrastructure: The Origin of Cell Organelles.* John H. Luft
1970. *Ecosystem Structure and Function,* Eugene P. Odum
1971. *The Biology of Behavior,* Bernard W. Agranoff

Ecosystem
Structure and Function

Proceedings of the Thirty-First
Annual Biology Colloquium

Edited by JOHN A. WIENS

OREGON STATE UNIVERSITY PRESS

Printed on Recycled Paper

© 1972 by the Oregon State University Press
ISBN: 0-87071-170-9   LCC: 52-19235
Printed in the United States of America

Preface

"Ecology," a term which once had a fairly precise meaning to a small closed group of scientists, has recently become part of the average citizen's vocabulary. Such diverse groups as detergent manufacturers and politicians have picked up the word (but unfortunately not always the concept) and now employ it liberally and with great fervency, but with different things in mind. This etymological change has occurred as man has become aware of the progressing state of environmental decay about him, of the wholesale disruption of the "balance of nature" by massive pollution of many sorts—as he has begun to realize the magnitude of the environmental crisis. Many feel that the solution to this crisis is within the technological power of western civilization. "After all, we put men on the moon," one hears. "Surely we can clean up the air and water." Further, ecologists are increasingly called upon as the "environmental engineers" who will provide the solutions to these problems.

The immediate response of ecologists, however, is that the problems are of a larger dimension than simply "cleaning up the air and water." Ecology deals with systems which are distressingly complex: *ecosystems*—groupings of populations of organisms and environmental factors which interact in time and space in complicated and often subtle and unpredictable manners. During the last decade or so ecologists have become increasingly aware of the vast complexity of these systems. With this understanding has emerged the realization that what we once thought were simple solutions to single problems have more often than not contributed to larger, global problems of ecosystem imbalance. Such imbalances defy simple solutions. Further, if the solutions are in fact technological, even employing a new ecological wisdom, they must involve a different view of achievement. In our culture, technological success is measured by growth, and growth invariably generates new problems (which, of course, require additional technology for their solution). What is needed, instead, is a view of technological achievement measured not in growth but in stability, the stability of man's ecosystem.

7

The roots of the environmental crisis are, of course, more than just technological, and involve more than ecosystem imbalance alone. Religious beliefs, ethics, sociological patterns, governmental philosophies, demography, economics, various behavioral and morphological vestiges of our primate ancestry, and a host of culturally fostered attitudes are also implicated and must be included in the search for solutions. But to continue to ignore man's relationship with nature and make political and economic decisions on environmental policy and resource management in the absence of ecological knowledge is senseless. It is urgent that our understanding of the structure and functioning of ecosystems be expanded and at the same time be introduced into policy decisions and the generation of environmental solutions.

The six papers and the discussions included in this volume obviously cannot adequately treat the entire complexity of ecosystems. The authors have examined various aspects of ecosystem structure and function, drawing heavily from their own work. Their presentations are individualistic, in the true spirit of a colloquium, but they share an awareness of the interrelatedness of ecosystem components and processes. Moreover, all produce generalizations or insights of considerable significance to the analysis and management of human ecosystems.

The colloquium would have been impossible without the financial support of the organizations listed in the Appendix. I am most grateful for the hard work of members of the local committee and the Oregon State University Chapter of the Ecological Society of America. I especially thank C. David McIntire, who flawlessly supervised many of the details of local arrangements. James D. Hall and William G. Pearcy provided editorial assistance, and the Department of Zoology at Oregon State University assisted materially in the preparation of the manuscript for publication.

JOHN A. WIENS

Contents

Ecosystem Theory in Relation to Man

EUGENE P. ODUM

Institute of Ecology
University of Georgia, Athens, Georgia 30601

THE CONCEPT OF THE ECOSYSTEM is not only the center of professional ecology today, but it is also the most relevant concept in terms of man's environmental problems. In the past two years the public has seized on the root meaning of ecology, namely *"oikos"* or "house," to broaden the subject beyond its previously rather narrow academic confines to include the "totality of man and environment," or the whole environmental house, as it were. We are witnessing what I have called a historic "attitude revolution" (Odum, 1969, 1970c) in the way people look at their environment for the very simple reason that for the first time in his short history man is faced with ultimate rather than merely local limitations. It will be well for all of us to keep this overriding simplicity in mind as we face the controversies, false starts, and backlashes that are bound to accompany man's attempts to put some negative feedback into the vicious spiral of uncontrolled growth and resource exploitation that has characterized the past several decades.

As recently as ten years ago the theory of the ecosystem was rather well understood but not in any way applied. The applied ecology of the 1960's consisted of managing components as more or less independent units. Thus we had forest management, wildlife management, water management, soil conservation, pest control, etc., but no ecosystem management and no applied human ecology. Practice has now caught up with theory! Controlled management of the human population together with the resources and the life support system on which it depends as a single, integrated unit now becomes the greatest, and certainly the most difficult, challenge ever faced by human society.

11

As we have seen, an "anthropocentric" definition of the ecosystem might read something as follows: Man as a part of, not apart from, a life-support system composed of the atmosphere, water, minerals, soil, plants, animals, and microorganisms that function together to keep the whole viable. If you would like a more formal definition, I can give you the one I am using in the third edition of *Fundamentals of Ecology* (Odum, 1971), namely: "Any unit including all of the organisms (i.e., the "community") in a given area interacting with the physical environment so that a flow of energy leads to a clearly defined trophic structure, biotic diversity, and material cycles (i.e., exchange of materials between living and non-living parts) within the the system is an *ecological system* or *ecosystem.*"

Historical Review of the Ecosystem Concept

Although the term ecosystem was first proposed by the British ecologist A. G. Tansley just 35 years ago (1935) the concept is by no means so recent. Allusions to the idea of the unity of organisms and environment (as well as the oneness of man and nature) can be found as far back in written history as one might care to look, and such an idea has been a basic part of many religions, although less so in Christian religions, as recently pointed out by historian Lynn White (1967). Anthropologists and geographers have long been concerned with the impact of man on his environment and early debated the question: To what extent has man's continuing trouble with deteriorated environments stemmed from the fact that human culture tends to develop independently of the natural environment? The Vermont prophet George Perkins Marsh wrote a classic treatise on this theme in 1864. He analyzed the causes of the decline of ancient civilizations and forecast a similar doom for modern ones unless man takes what we would call today an "ecosystematic" view of man and nature. In the late 1800's biologists began to write essays on the unity of nature, interestingly enough in a parallel manner in German, English, and Russian languages. Thus Karl Mobius in 1877 wrote about the community of organisms in an oyster reef as a "biocoenosis," while in 1887 the American S. A. Forbes wrote his classic essay on "The Lake as a Microcosm." The Russian pioneering ecologist V. V. Dokuchaev (1846-1903) and his disciple G. F. Morozov (who specialized in forest ecology) placed great emphasis on the concept of "biocoenosis," a term later expanded to geobiocoenosis (or biogeocoenosis) (*see* Sukachev, 1944), which can be considered a synonym of the word "ecosystem."

No one has expressed the relevance of the ecosystem concept to man better than Aldo Leopold in his essays on the land ethic. In 1933 he wrote: "Christianity tries to integrate the individual to society, democracy to integrate social organization to the individual. There is yet no ethic dealing with man's relation to the land" which is "still strictly economic, entailing privileges but not obligations." Thus, man is continually striving, with but partial success so far, to establish ethical relationships between man and man, man and government, and, now, man and environment. Without the latter what little progress has been made with the other two ethics will surely be lost. In the context of the 1970 scene Garrett Hardin (1968) says it in another way when he points out that technology alone will not solve the population and pollution dilemmas; ethical and legal constraints are also necessary. Environmental science is now being called upon to help determine a realistic level of human population density and rate of use of resources and power that are optimum in terms of the quality of human life, in order that "societal feedback" can be applied before there are serious overshoots. This requires diligent study of ecosystems, and ultimately, a judgment on the carrying capacity of the biosphere. If studies of natural populations have any bearing on the problem, we can be quite certain that the optimum density in terms of the individual's options for liberty and the pursuit of happiness is something less than the maximum number that can be sustained at a subsistence level, as so many domestic "animals" in a polluted feed lot!

My advanced ecology class recently attempted to determine what might be the optimum population for the state of Georgia on the assumption that someday the state would have to have a balanced resource input-output (i.e., live within its own resources). On the basis of a per capita approach to land use the tentative conclusion was that a density of one person per 5 acres (2 hectares) represented the upper limit for an optimum population size when the space requirements for quality (i.e., high protein) food production, domestic animals, outdoor recreation, waste treatment, and pollution-free living space were all fully considered. Anything less than 5 acres of life support and resource space per capita, it was concluded, would result in a reduction in the individual person's options for freedom and the pursuit of happiness, and accordingly, a rapid loss in environmental quality. Since the 1970 per capita density of Georgia is 1 in 8 acres, and for the United States as a whole, 1 in 10 acres, no more than double the present U. S. population could be considered optimum according to this type of analysis. This would mean that we have about 30 years to level off population growth. The study also suggested that permanent zoning of at least one-third of land and freshwater areas (plus all estuarine and marine zones) as "open space" in urbanized areas would go a long way towards preventing the overpopulation, overdevelopment, and social decay that is now so evident in many parts of the world today.

The results of this preliminary study have been published (E. P. Odum, 1970b; *see also* the provocative symposium on "The Optimum Population for Britian" edited by Taylor, 1970).

The Two Approaches to Ecosystem Study

G. Evelyn Hutchinson in his 1964 essay, "The Lacustrine Microcosm Reconsidered," contrasts the two longstanding ways ecologists attempt to study lakes or other large ecosystems of the real world. Hutchinson cites E. A. Birge's 1915 work on heat budgets of lakes as pioneering the *holological* (from *holos* = whole) or *holistic* approach in which the whole ecosystem is treated as a "black box" (i.e., a unit whose function may be evaluated without specifying the internal contents) with emphasis on inputs and outputs, and he contrasts this with the *merological* (from *meros* = part) approach of Forbes in which "we discourse on parts of the system and try to build up the whole from them." Each procedure has obvious advantages and disadvantages and each leads to different kinds of application in terms of solving problems. Unfortunately, there is something of a "credibility gap" between the two approaches.

As would be expected, the merological approach has dominated the thinking of the biologist-ecologist who is species-oriented, while the physicist-ecologist and engineer prefer the "black box" approach. Most of all, man's environmental crisis has speeded up the application of systems analysis to ecology. The formalized or mathematical model approach to populations, communities, and ecosystems has come to be known as *systems ecology*. It is rapidly becoming a major science in its own right, for two reasons: (1) Extremely powerful new formal tools are now available in terms of mathematical theory, cybernetics, electronic data processing, etc.; and (2) formal simplication of complex ecosystems provides the best hope for solutions of man's environmental problems that can no longer be trusted to trial-and-error or one-problem-one-solution procedures that have been chiefly relied on in the past.

Again we see the contrast between merological and holological approaches in that there are systems ecologists who start at the population or other component level and "model up," and those who start with the whole and "model down." The same dichotomy is evident in the very rewarding studies of experimental laboratory ecosystems. One class of microecosystems can be called "derived" systems, because they are established by multiple seeding from nature, in contrast to "defined" microcosms which are built up from previously isolated pure cultures. Theoretically, at least, the approaches are applicable to efforts to devise life-support systems for space travel. In fact, one of the best ways to visualize the ecosystem concept for students and laymen is to consider space travel, because when man leaves the biosphere he must take with him a sharply delimited enclosed environment that will supply all vital needs with solar energy as the only usable input from the surrounding very hostile space environment. For journeys of a few weeks (such as to the moon and back) man does not need a regenerative ecosystem, since sufficient oxygen and food can be stored while CO_2 and other waste products can be fixed or detoxified for short periods of time. For long journeys man must engineer himself into a complete ecosystem that includes the means of recycling materials and balancing production, consumption, and decomposition by biotic components or their mechanical substitutes.

In a very real sense the problems of man's survival in an artificial space craft are the same as the problems involved in his continued survival on the earth space ship. For example, detection and control of air and water pollution, adequate quantity and nutritional quality of food, what to do with accumulated toxic wastes and garbage, and the social problems created by reduced living space are common concerns of cities and spacecrafts. For this reason the ecologist would urge that national and international space programs now turn their attention to the study and monitoring of our spaceship earth. As was the case with Apollo 13, survival becomes the mission when the limits of carrying capacity are approached.

The Components of the Ecosystem

From the energy or trophic standpoint an ecosystem has two components which are usually partially separated in space and time, namely, an *autotrophic component* (*autotrophic* = self nourishing) in which fixation of light energy, use of simple inorganic substances, and the buildup of complex substances predominate; and secondly, a *heterotrophic component* (*heterotrophic* = other-nourishing) in which utilization, rearrangement, and decomposition of complex materials predominate. As viewed from the side (cross section) ecosystems consist of an upper "green belt" which receives incoming solar energy and overlaps, or interdigitates, with a lower "brown belt" where organic matter accumulates and decomposes in soils and sediments.

It is convenient for the purposes of first-order analysis and modeling to recognize six structural components and six processes as comprising the ecosystem:

A. Components

1. *Inorganic substances* (C, N, CO_2, H_2O, etc.) involved in material cycles.

2. *Organic compounds* (proteins, carbohydrates, lipids, humic substances, etc.) that link biotic and abiotic components.

3. *Climate regime* (temperature, rainfall, etc.).

4. *Autotrophs* or *producers,* largely green plants able to manufacture food from simple substances.

5. *Phagotrophs* (*phago* = to eat) or *macro-consumers,* heterotrophic organisms, largely animals which ingest other organisms or particulate organic matter.

6. *Saprotrophs* (*sapro* = to decompose) or *micro-consumers* (also called osmotrophs), heterotrophic organisms, chiefly bacteria, fungi, and some protozoa, that break down complex compounds, absorb some of the decomposition products, and release inorganic substances usable by the autotrophs together with organic residues which may provide energy sources or which may be inhibitory, stimulatory, or regulatory to other biotic components of the ecosystem. (Another useful division for heterotrophs: *biophages* = organisms that feed on other living organisms; *saprophages* = organisms that feed on dead organic matter.)

B. Processes

1. *Energy flow circuits.*

2. *Food chains* (trophic relationships).

3. *Diversity patterns* in time and space.

4. *Nutrient* (*biogeochemical*) *cycles.*

5. *Development and evolution.*

6. *Control* (cybernetics).

Subdivision of the ecosystem into these six "components" and six "processes," as with most classifications, is arbitrary but convenient, since the former emphasize structure and the latter function. From the holistic viewpoint, of course, components are operationally inseparable. While different methods are often required to delineate structure on the one hand and to measure rates of function on the other, the ultimate goal of study at any level of organization is to understand the relationship between structure and function. It is not feasible to go into any detailed discussion of these component-processes in this brief introduction, but we can list a few key principles that are especially relevant to human ecology. Figure 1 is a schematic diagram that may be useful in picturing the basic arrangement and functional linkage of ecosystem components.

1. The living (items A 4-6 above) and non-living (A 1-3) parts of ecosystems are so interwoven into the fabric of nature that it is difficult to separate them; hence operational classifications (B 1-6) do

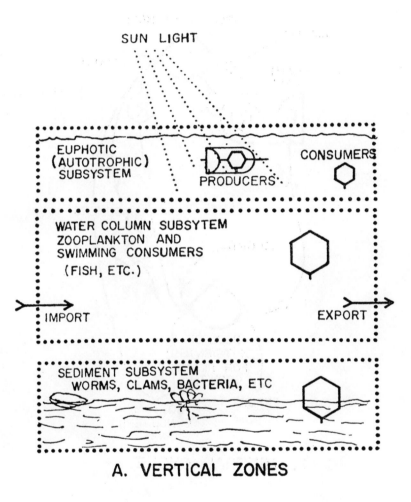

A. VERTICAL ZONES

FIGURE 1. Three aspects of the structure and function of ecosystems as illustrated by an estuarine system. **A.** Vertical zonation with photosynthetic production above (autotrophic stratum) and most of the

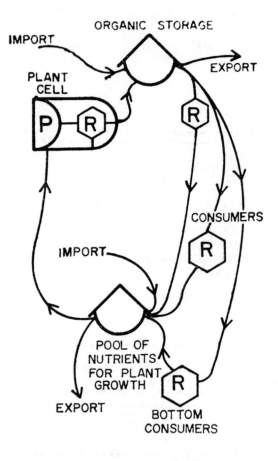

B. MINERAL CYCLE

respiration and decomposition below (heterotrophic stratum). **B.** Material cycle with circulation of plant nutrients upward and organic

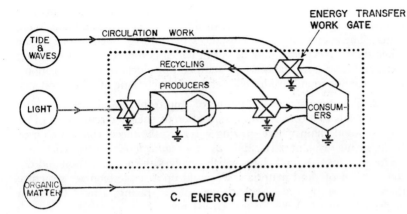

C. ENERGY FLOW

matter food downward. **C.** Energy flow circuit diagram showing three sources of energy input into the system. The bullet-shaped modules represent producers with their double metabolism, that is P (production) and R (respiration). The hexagons are populations of consumers which have storage, self maintenance, and reproduction. The storage bins represent nutrient pools in and out of which move nitrogen, phosphorus, and other vital substances. In diagrams B and C the lines represent the "invisible wires of nature" that link the components into a functional network. In diagram C the "ground" symbols (i.e., arrow into the heat sink) indicate where energy is dispersed and no longer available in the food chain. The circles represent energy inputs. The work gate symbols (large X) indicate where a flow of work energy along one pathway assists a second flow to pass over energy barriers. Note that some of the lines of flow loop back from "downstream" energy sources to "upstream" inflows serving various roles there, including control functions (saprotrophs controlling photosynthesis by controlling the rate of mineral regeneration, for example). The diagram (C) also shows how auxillary energy of the tide (energy subsidy) assists in recycling of nutrients from consumer to producer and speeding up the movement of plant food to the consumer. Reducing tidal flow by diking the estuary will reduce the productivity just as surely as cutting out some of the light. Stress such as pollution or harvest can be shown in such circuit models by adding circles enclosing negative signs linked with appropriate heat sinks to show where energy is diverted away from the ecosystem. Both subsidies (+) and stresses (−) can be quantified in terms of calories aded or diverted per unit of time and space. (From E. P. Odum, 1971, after H. T. Odum, Copeland, and McMahan, 1969.)

not make a sharp distinction between biotic and abiotic. Elements and compounds are in a constant state of flux between living and non-living states. There are very few substances that are confined to one or the other state. Exceptions may be ATP, which is found only inside living cells, and humic substances (resistant end products of decomposition) which are not found inside cells yet are characteristic of all ecosystems.

2. The time-space separation of autotrophic and heterotrophic activity leads to a convenient classification of energy circuits into (1) a *grazing* food chain where grazing refers to the direct consumption of living plants or plant parts, and (2) an *organic detritus* (from *deterere* = to wear away) food chain which involves the accumulation and decomposition of dead materials. To build up a stable biomass structure there must be negative feedback control of grazing, a need too often neglected in man's domesticated ecosystems.

3. As is well known, available energy declines with each step in the food chain (so a system can support more herbivores than carnivores; if man wants to keep his meat-eating option open there will have to be fewer people supported by a given food base). On the other hand, materials often become concentrated with each step in the food chain. Failure to anticipate possible "biological magnification" of pollutants, such as DDT or long-lived radionuclides, is causing serious problems in man's environment.

4. It is becoming increasingly evident that high biological productivity (in terms of calories per unit area) in both natural and agricultural ecosystems is almost always achieved with the aid of *energy subsidies* from outside the system that reduce the cost of maintenance (thus diverting more energy to production). Energy subsidies take the form of wind and rain in a rain forest, tidal energy in an estuary (*see* Fig. 1), or fuel, animal, or human work energy used in the cultivation of a crop. In comparing productivity of different systems it is important to consider the *complete budget*—not just sunlight input.

5. Likewise it is increasingly evident that both *harvest* and *pollution are stresses* which reduce the energy available for self-maintenance. Man must be aware that he will have to pay the costs of added antithermal maintenance, or "disorder pumpout" as H. T. Odum (1967, 1970) calls it. It is a dangerous strategy to try to force too much productivity or yield from the landscape (as is being attempted in the so-called "green revolution") because very serious "ecological backlashes" can occur. These may result from: (1) pollution caused by heavy use of fertilizers and insecticides and the consumption of fossil fuels; (2) unstable or oscillating conditions created by one-crop systems; (3) vulnerability of plants to disease because their self-protection me-

chanisms have been "selected out" in favor of yield; and (4) social disorder created by a rapid shift of rural people to cities that are not prepared to house or employ them. (The tragedy here is that industrialized agriculture *can* result in increased food per acre but it can also widen the gap between rich and poor so that there are increasing numbers of people unable to buy the food!)

6. While we generally think of production and decomposition as being balanced in the biosphere as a whole, the truth is that this balance has never been exact but has fluctuated from time to time in geological history. Through the long haul of evolutionary history, production has slightly exceeded decomposition so that a highly oxygenic atmosphere has replaced the original reducing atmosphere of the earth. Man, of course, is tending to reverse this trend by increasing decomposition (burning of fuels, etc.) at the expense of production. The most immediate problem is created by the increase in atmospheric CO_2, since relatively small changes in concentration can have large effects on the heat budget of the earth.

7. Ecological studies indicate that diversity is directly correlated with stability and perhaps inversely correlated with productivity, at least in many situations. It could well be that the preservation of diversity in the ecosystem is important for man since variety may be a necessity, not just the spice of life!

8. At the population level it is now clear that the growth form of the human population will not conform to the simple sigmoid or logistic model since there will always be a long time lag in the effects of crowding, pollution, and overexploitation of resources. Growth will not "automatically" level off as do populations of yeasts in a confined vessel where individuals are *immediately* affected by their waste products. Instead, the human population will clearly overshoot some vital resource unless man can "anticipate" the effects of overpopulation and reduce growth rates *before* the deleterious effects of crowding are actually felt. Intelligent reasoning behavior seems now to be the only means to accomplish this, as I emphasized at the beginning of this article.

9. Some of the most important "breakthroughs" in ecology are in the area of biogeochemical cycling. Since "recycle" of water and minerals must become a major goal of human society, the recycle pathways in nature are of great interest; there seem to be at least four major ones which vary in importance in different kinds of ecosystems: (1) recycle via microbial decomposition of detritus; (2) recycle via animal excretion; (3) direct recycle from plant back to plant via symbiotic microorganisms such as mycorrhizae associated with roots;

and (4) autolysis, or chemical recycle, with no organism involved. Pathway 3 seems to be especially important in the humid tropics, which suggests that tropical agriculture might be redesigned to include food plants with mycorrhizae.

10. The principles inherent in limiting factor analysis and in human ecology can be combined to formulate the following tentative overview: In an industrialized society energy (power, food) is not likely to be limiting, *but the pollution consequences of the use of energy and exploitation of resources are limiting.* Thus, pollution can be considered the limiting factor for industrialized man—which may be fortunate since pollution is so "visible" that it can force us to use that reasoning power which is supposed to be our special attribute.

Ecosystem Development

Principles having to do with the development of ecosystems, that is, ecological succession, are among the most relevant in view of man's present situation. I have recently reviewed this subject (Odum, 1970a); accordingly a brief summary will suffice here.

In broad view ecosystems develop through a rapid growth stage that leads to some kind of maturity or steady state (climax), usually an oscillating steady state. The early successional growth stage is characterized by a high production/respiration (P/R) ratio, high yields (net production), short food chains, low diversity, small size of organisms, open nutrient cycles, and a lack of stability. In contrast, mature stages have a high biomass/respiration (B/R) ratio, complex food webs, low net production, and high diversity and stability. In other words, major energy flow shifts from production to maintenance (respiration).

The general relevance of the development sequence to land-use planning can be emphasized by the following "mini-model" that contrasts in very general terms young and mature ecosystems:

Young	*Mature*
Production	Protection
Growth	Stability
Quantity	Quality

It is mathematically impossible to obtain a maximum for more than one thing at a time, so one can not have both extremes at the same time and place. Since all six characteristics are desirable in the aggregate, two possible solutions to the dilemma suggest themselves. We can compromise so as to provide moderate quality and moderate yield on all the landscape, or we can plan to compartmentalize the landscape so as to maintain simultaneously highly productive and predominantly protective types as separate units subjected to different management strategies. If ecosystem development theory is valid and applicable to land-use planning (total zoning), then the so-called multiple-use strategy, about which we hear so much, will work only through one or both of these cases, because in most cases projected multiple uses conflict with one another.

Literature Cited

Birge, E. A. 1915. The heat budgets of American and European lakes. Trans. Wis. Acad. Arts. Lett., *18*:166-213.

Dokuchaev, V. V. "Teaching about the zones of nature." (In Russian.) Reprinted 1948. Moscow.

Forbes, S. A. 1887. The lake as a microcosm. Bull. Sc. H. Peoria. Reprinted in: Ill. Nat. Hist. Surv. Bull., *15*:537-550. 1925.

Hardin, G. 1968. The tragedy of the commons. Science, *162*:1243-1248.

Hutchinson, G. E. 1964. The lacustrine microcosm reconsidered. Amer. Sci., *52*:331-341.

Leopold, A. 1933. The conservation ethic. Jour. Forestry, *31*:634-643.

Marsh, G. P. 1864. *Man and Nature: Or Physical Geography as Modified by Human Action.* Reprinted 1965 by Harvard Univ. Press, Cambridge.

Mobius, K. 1877. Die Auster und die Austernwirtschaft. Berlin. Transl. 1880. Rept. U. S. Fish Comm., *1880*:683-751.

Odum, E. P. 1969. The attitude lag. BioScience, *19*:403.

Odum, E. P. 1970a. The strategy of ecosystem development. Science, *164*:262-270.

Odum, E. P. 1970b. Optimum population and environment: a Georgia microcosm. Current History, *58*:355-359.

Odum, E. P. 1970c. The attitude revolution. In: *The Crisis of Survival*, pp. 9-15. Scott, Foresman Co., Glenview, Ill.

Odum, E. P. 1971. *Fundamentals of Ecology*. 3rd edition. W. B. Saunders, Philadelphia.

Odum, H. T. 1967. Biological circuits and the marine systems of Texas. In T. A. Olson and F. J. Burgess [eds.], *Pollution and Marine Ecology*, pp. 99-157. John Wiley and Sons, N. Y.

Odum, H. T. 1970. *Environment, Power and Society*. John Wiley and Sons, N. Y.

Odum, H. T., B. J. Copeland, and E. A. McMahan. 1969. Coastal ecological systems of the United States. A report to the Federal Water Pollution Control Administration, Marine Sci. Inst., Univ. Texas. Mimeograph.

Sukachev, V. N. 1944. On principle of genetic classification in biocoenology. Translated and condensed by F. Raney and R. Daubenmire. Ecology, *39*:364-367. (*See also* Silva Fennica, *105*:94. 1960.)

Tansley, A. G. 1935. The use and abuse of vegetational concepts and terms. Ecology, *16*:284-307.

Taylor, L. R. (Ed.). 1970. *The Optimum Population for Britain.* Academic Press, N. Y.

White, L. 1967. The historical roots of our ecological crisis. Science, *155*:1203-1207.

Discussion

QUESTION: The "ecology movement" has quite effectively drawn attention to the multitude of environmental problems which confront us, but ecologists seem often to be unaware that we can't do everything, that we must establish some environmental priorities. How do you suggest we should approach problems such as, say, pollution?

DR. ODUM: We have two priorities in regard to pollution abatement that I think we all recognize as major concerns of the "ecology movement" in this country. One priority is to correct mistakes of the past such as attempting to clean up bad pollution and outlawing the dumping of dangerous toxic substances into the environment. A second priority is to prevent further deterioration of the quality of environment. The latter priority, it seems to me, should now receive the greater attention, especially since it is economically more feasible to prevent rather than to clean up. After all, we can live with present levels of pollution, but we will suffer very grievously if pollution and the accompanying waste of resources continues to increase, and sooner or later it will become economically impossible to pay for restoration or recovery of wasted resources. Prevention requires (1) that we "power down" and "recycle" the use of resources and (2) that waste management and land-use planning be given first, not last, consideration in all future industrial, agricultural, and urban development. I indicated some approaches to this in my paper, and I hope that after all the other papers in this symposium have been presented we can come back to future strategies in our final discussion.

Nutrient Cycling in Ecosystems

GENE E. LIKENS*
*Section of Ecology and Systematics
Cornell University, Ithaca, New York 14850*
and
F. HERBERT BORMANN
*School of Forestry
Yale University, New Haven, Connecticut 06511*

THE WISE MANAGEMENT of our natural resources depends upon a sound understanding of the structure and function of ecological systems. To date, the narrow approaches such as specialized agricultural and industrial strategies, designed to maximize production of food, power, and other products, have dominated our management of natural resources and invariably have led to imbalances and instabilities in ecosystems. When resource management is based on an understanding of the ecosystem's interconnections and interactions, the "hidden costs of narrow management strategies" become part of the overall accounting, and further environmental deterioration and "confrontation in the cornfield" can be avoided (Cantlon, 1969). On the hopeful side, agricultural scientists are starting to talk about optimizing production rather than maximizing it.

In the past, when natural resources were plentiful in relation to man's wants and abilities for utilization, the function of natural ecosystems was considered much less important than their structure. Forests were to be cleared, rivers to be tamed, and wild animals to be conquered. Now in the face of man's exploding population and dwindling resource base, his very survival may depend on an accurate knowledge of ecosystem function, i.e., maintaining the continuous flow of energy and nutrients vital to the existence of ecological systems and life

---

*This paper was delivered at the Colloquium by Dr. Likens.

itself. Within an ecosystem, substances (cations, anions, molecules) are continually withdrawn from the abiotic reservoir, utilized by and circulated through the biotic portion, and returned in one form or another to the abiotic reservoir. Leopold (1949: 104-108) provides an elegant description of this in his "Odyssey" of atom X. "Time, to an atom locked in a rock, does not pass. The break came when a bur-oak root nosed down a crack and began prying and sucking. In the flash of a century the rock decayed, and X was pulled out and up into the world of living things. He helped to build a flower, which became an acorn, which fattened a deer, which fed an Indian, all in a single year." Leopold follows the itinerant atom's Odyssey through various pathways in the abiotic and biotic environment, before it ends ultimately in the sea; then summarizes with, "An atom at large in the biota is too free to know freedom; an atom back in the sea has forgotten it. For every atom lost to the sea, the prairie pulls another out of the decaying rocks. The only certain truth is that its creatures must suck hard, live fast, and die often, lest its losses exceed its gains."

The quantitative study of odysseys of nutrient elements has become one of the major thrusts of current ecological research. It is now clear that an understanding of nutrient cycles is one of the touchstones to understanding the welfare of man. For example, there is a finite quantity of certain essential minerals or nutrients (e.g., phosphorus) available to land plants in the biosphere. It is estimated that some 8 million metric tons of PO_4 are transported to the sea each year and of this, only a small percentage is recycled (returned) annually to the land (Cole, 1958).

A Model for the Study of Nutrient Cycles

Terrestrial Ecosystem Model

For the purposes of our discussions we will focus on the deciduous forest biome, since we are most familiar with its structure and function. The vertical boundaries of the functional ecosystem are determined by biological utilization of energy and nutrients. Lateral boundaries are arbitrarily set by the investigator, but may coincide with natural boundaries such as the edge of a forest. The continuous flow of nutrients and other chemicals through these ecosystem boundaries may be visualized as meteorologic, geologic, and biologic inputs and outputs (Fig. 1). Meteorologic inputs and outputs consist of nutrients as wind-borne particulate matter, dissolved substances in rain and snow, or gases (e.g., CO_2). Geologic nutrient flux includes dissolved and particulate matter transported by surface and subsurface drainage, and the mass

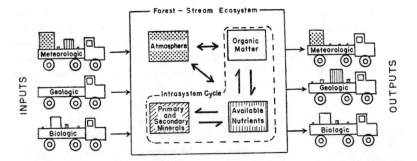

FIGURE 1. A diagrammatic model for nutrient cycling in forest-stream ecosystems. Inputs and outputs to the ecosystem are moved by meteorologic, geologic, or biologic vectors. "Cargos" are arranged from left to right: atmosphere, organic matter, available nutrients, primary and secondary minerals. Major sites of accumulation and exchange pathways within the ecosystem are shown.

movement of colluvial materials. Biologic flux results when nutrients gathered by animals in one ecosystem are deposited in another (e.g., local exchange of fecal matter or mass migrations). Thus these input-output categories are defined as vectors or "vehicles" for nutrient transport, rather than sources of nutrients; i.e., a leaf blown into an ecosystem would represent meteorologic input rather than biologic input.

Within the ecosystem, the nutrients may be thought of as occurring in any one of four basic compartments: (1) atmosphere, (2) living and dead organic matter, (3) available nutrients, and (4) primary and secondary minerals (soil and rock). The atmospheric compartment includes all elements in gaseous form both above and below ground. Available nutrients are ions that are absorbed in the clay-humus complex or dissolved in the soil solution. The organic compartment includes all nutrients incorporated in living and dead biomass. The primary and secondary minerals contain nutrients that comprise the inorganic soil and rock portions of the ecosystem.

The biogeochemical flux of elements involves an exchange between the various compartments of the ecosystem. Available nutrients and gaseous nutrients may be taken up and assimilated by the vegetation and microorganisms, some passed on to heterotrophic consumers, and then made available again through respiration, biological decomposition, and/or leaching from living and dead organic matter. Insoluble primary and secondary minerals may be converted to soluble available nutrients through the general process of weathering; soluble nutrients may be redeposited as secondary minerals. Because nutrients with a sedimentary biogeochemical cycle (no prominent gaseous phase, Odum, 1959) are continually recycled within the boundaries of the ecosystem between the available nutrient, organic matter, and primary and secondary mineral compartments, they tend to form an intrasystem cycle (Bormann and Likens, 1967).

This model allows for development (biological succession) or degradation of the ecosystem. This is expressed in several ways: (1) by an increase or decrease in the functional volume of the ecosystem, (2) by alterations in the flux rate between compartments, (3) by changes in the size of the compartments, and (4) by a change in output relationships. Consider, for example, the developmental changes that would occur during xerarch succession initiated on a bare granitic rock (Figs. 2 and 3). Initially the volume of the ecosystem would be quite small, essentially the volume occupied by mosses and lichens and a thin layer of soil. The organic compartment would contain only the biomass of a few small plants, animals, and microorganisms, plus a small amount of dead organic matter. Available nutrients would be limited to the few held on a very limited soil-humus exchange complex, and secondary minerals (clay) would be almost nonexistent. Consistent with these small compartments, nutrient uptake, organic decomposition, weathering, and formations of secondary minerals would be relatively small. There would be a relatively large liquid output, since most of the water would run off during a rain, and a very small evapotranspirational output contributing to the hydrologic budget for this system.

With development of the ecosystem during xerarch succession, its volume would increase as its vertical boundaries extended into the atmosphere (i.e., with taller plants) and into the decaying rock below. Components of the various compartments, living biomass, dead biomass including humus, available nutrients, and secondary minerals (e.g., clay) would increase in amount, and flux rates between compartments would increase. The proportion of water loss as liquid hydrologic output would be diminished, while evapotranspiration would increase. Thus, the ecosystem model provides a framework whereby we can consider not only structure and function, but also the development or degradation of an ecosystem. It allows us to assemble and interrelate an extraordinarily diverse array of biogeochemical data into one coherent system.

Forest-Stream Ecosystem—Small Watershed Approach

It is obvious that quantification of all the components and interactions of this model is difficult. Nutrients are constantly being added

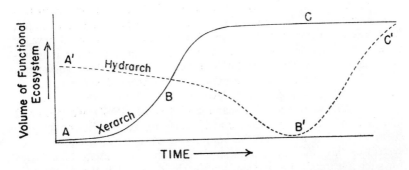

FIGURE 2.   Volumetric relationships in xerarch (rock) and hydrarch (lake) development. **A** = bare rock, **B** = herb and shrub stage, **C** = mature mesic forest and soil, **A′** = open water, **B′** = sedge meadow on lake sediments and **C′** = swamp forest. The ultimate volumetric development is based upon a number of environmental factors.

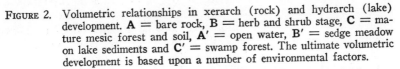

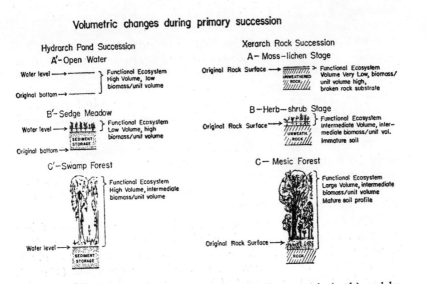

FIGURE 3.   Diagrammatic volumetric relationships for xerarch (rock) and hydrarch (lake) development.

and removed from the ecosystem, and quantitative measurement of this flux may be virtually impossible for some ecosystems. For example, how does one measure nutrient input or output in sheet or rill flow, or in water lost as deep seepage? Without such measurements it is impossible to construct quantitative nutrient budgets for ecosystems. Hence one is forced into dealing with many aspects of ecosystem nutrient relationships on a qualitative level.

Several years ago (Bormann and Likens, 1967) we proposed that it is conceptually useful and simplifying for studies of biogeochemical cycling to define the boundaries of a deciduous forest ecosystem as the limits of a watershed. Therefore the lateral boundaries for such a watershed ecosystem are clearly identified as the topographic divides. Nutrient inputs then are restricted to meteorologic and biologic vectors since by definition there can be no transfer of geologic inputs between adjacent watersheds. Furthermore, if the watershed ecosystem (1) is a part of a larger, homogeneous biotic and geologic unit, (2) has an impermeable geologic substrate, and (3) is characterized by relatively humid conditions, the input-output budget for nongaseous nutrients may be simply determined from the difference between the meteorologic input (dissolved substances and particulate matter in rain and snow) and the geologic output (dissolved substances and particulate matter in drainage waters). The amount and chemical quality of precipitation can be measured with a series of rain-gauging stations, and because of the impermeable substrate, the quantity and quality of all drainage water also can be accurately measured at a gauging station. Meteorologic output (dust, etc.) is negligible from this humid ecosystem, and it also is reasonable to assume that biological input balances biological output except where man produces a geographic distortion by hunting, application of fertilizer, etc. The size of the input and output "cargos" in Figure 1 are proportional to the relative magnitudes involved in the northern hardwoods forest watershed ecosystem.

To understand the utility of the small watershed technique it might be useful to examine the potential pathways for calcium as it flows through a deciduous forest ecosystem. Calcium has a sedimentary cycle, and consequently a quantitative input-output budget can be obtained by the watershed approach.

The meteorologic vectors, wind and precipitation, dump calcium into the ecosystem. Calcium in wind-borne organic matter (such as leaves) would contribute to or detract from the calcium already in the organic matter compartment of the intrasystem cycle (Fig. 1). Meteorologic input and output of inorganic particulate matter containing calcium (e.g., soil or road dust) would contribute to or delete from the amount of calcium in the primary and secondary mineral compartment, and ionic calcium in rain and snow would contribute, at least initially, to the available nutrient compartment. It should be noted at this point that although nutrients are added to a specific compartment, they then become a part of the intrasystem cycle and are subject to transformations and exchanges. Since evaporational and transpirational water outputs would be essentially free of ionic calcium and in humid forest ecosystems the meteorologic input and output of calcium in particulate matter is considered very minimal, the meteorologic flux of calcium for the watershed ecosystem can be estimated quantitatively from collections and chemical analysis of precipitation.

There would be no geologic input of calcium to this watershed ecosystem since gravity powered geologic vectors cannot cross the watershed divides. However, large amounts of calcium would be lost via geologic vectors, first, as dissolved calcium from the available nutrients compartment, second, through erosion and transport by drainage waters of calcium contained in primary and secondary minerals, third, by transport of dissolved and particulate organic matter containing calcium in the drainage stream, and finally from colluvial movement of primary and secondary minerals downslope. Each of these quantities can be measured with the aid of a weir anchored to the bedrock at the mouth of the watershed.

Biological input of calcium to the organic matter and available nutrients compartments would result if a deer that fed entirely elsewhere walked into the ecosystem and defecated. Likewise output would occur if the deer fed within the ecosystem and then walked outside to defecate. Very small amounts of primary and secondary minerals (and possibly some gases) might be fluxed through the ecosystem in this manner also. Larger inputs of potentially available nutrients as well as some organic matter and primary and secondary minerals would result from an intentional application of fertilizer by man. Assuming the watershed is part of a larger, more or less homogenous ecologic unit, biologic input and output would tend to cancel.

Given these patterns, the calcium cycle for the ecosystem could be characterized in kg/ha by (1) the input of calcium in precipitation, (2) the gross loss in drainage waters, (3) the net loss (output-input), (4) quantities of calcium in the various compartments of the ecosystem, and (5) flux rates for uptake, decomposition, and leaching. As will be discussed later, data on net losses may be used to calculate weathering rates.

Lake Ecosystem Model

Odum (1962) suggests that the supporting matrix is the most important structural difference between terrestrial and aquatic ecosystems. That is, the superstructure and ecological classification of a terrestrial ecosystem are determined by the biology (usually vegetation), whereas the aquatic ecosystem is structured on a physical basis, the water. If most of the air were to be instantaneously removed from a forest ecosystem, it still would appear about the same. The forest biomass maintains the structure of the ecosystem. However, if all of the water were removed from a lake, it no longer would be (or appear to be) a lake. Conversely if the trees were removed from a forest it no longer would be (or look like) a forest, but if all of the plankton were removed from a lake it still would look like a lake.

Many years ago Forbes (1887) discerned and assembled many of the interrelated components of a lake ecosystem and thus made a historic contribution to our understanding of lake ecology. However, his concept of the lake as a microcosm (the lake "forms a little world within itself—a microcosm," p. 537) may have focused attention for many years on the lake as a self-sustaining unit without due consideration of its vital interactions with the surrounding watershed.

Lateral boundaries of a functional lake ecosystem are clearly delimited by the shoreline: the vertical boundaries are the surface of the water and the maximum sediment depth utilized by organisms.

External hydrologic relationships for a lake ecosystem are basic to an evaluation of its biogeochemistry. Inputs of water occur as: (1) precipitation falling directly on the lake's surface, (2) drainage of surface water into the lake, (3) seepage of ground water through the walls of the basin, and (4) discharge of lascustrine springs. All of these sources of water may add nutrients as dissolved or suspended impurities. Water losses occur through: (1) evaporation and transpiration, (2) surface effluents, (3) seepage through the basin wall, and (4) discrete subsurface flows.

Many small bog lakes may have neither surface inlet nor outlet drainage, the water balance being maintained by seepage of ground water, precipitation, and evaporation. Birge and Juday (1934) referred to such lakes as seepage lakes and distinguished them from drainage lakes. However, Broughton's data (1941) indicated that the basins of many bog lakes are effectively sealed against seepage, except for some surficial exchange. Moreover, Hutchinson (1957) believes that the hydrologic importance of subsurface discharge from springs is generally overrated. However, for certain rare crenogenic meromictic lakes (Hutchinson, 1957) the input of dissolved chemicals by sublacustrine springs may be enormous. Thus it would appear that the

NUTRIENT INPUTS

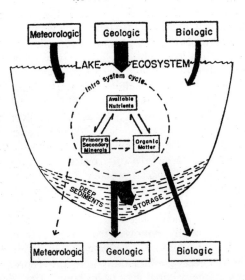

FIGURE 4. A diagrammatic model for nutrient cycling in a lake ecosystem.

principal hydrologic factors for most lakes are precipitation, evapo-transpiration, and surface drainage.

Nutrient inputs and outputs for the lake ecosystem as a whole can be categorized as was done in the terrestrial ecosystem into geologic, meteorologic, and biologic vectors (Fig. 4). The geologic input to a lake ecosystem is derived primarily as the geologic output from the surrounding terrestrial (forest-stream) ecosystem and therefore represents one of the important linkages between ecosystems in the biosphere. If the lake's watershed is utilized for agricultural or urban purposes, then the linkage becomes ecologically more critical. Lake Erie represents probably the best (sad?) example of the overwhelming effects of unnatural "geologic" inputs. From a pragmatic standpoint, it is necessary to study the terrestrial watershed to obtain an accurate prediction of geologic inputs into the lake. Meteorologic inputs are similar to those for terrestrial ecosystems. Biological inputs and outputs may be larger and more directional, relative to the total budgets, than for terrestrial ecosystems. For example terrestrially based predators (e.g., kingfishers) feeding on aquatic fauna, or spawning migrations of salmon, or insect emergences (e.g., mayflies) may contribute to significant removal or addition of nutrients to a lake ecosystem. Some of these fluxes may be equal over a period of time longer than a year.

Meteorologic nutrient outputs from a lake ecosystem are very small. Spray or aerosols may be generated from large lakes on windy days, but the nutrient content is relatively low. There may be appreciable gaseous flux of nutrients such as carbon dioxide and others (methane, nitrogen, hydrogen sulfide, etc.), particularly from some shallow anaerobic lake ecosystems.

Geologic outputs occur through losses of dissolved or particulate matter in any drainage water from lakes, although the particulate matter losses would usually be relatively small since the lake acts as a settling basin for suspended materials. In this regard, geologic outputs from a lake ecosystem represent a special problem in relation to the concept of ecosystem boundaries. Sediments are constantly accumulating in the lake basin. Since the living biota do not penetrate these sediments to great depths, those sediments below a relatively shallow crust of the lake bottom are essentially removed from further cycling within the lake and must be considered as geologic output from the functional ecosystem. These deeper sediments potentially represent long-term storage in terms of the original ecosystem boundaries, since conditions governing the ecology of the lake may change and some deep sediments may be brought back into nutrient circulation. They then would be considered as nutrient input. Such "stored" nutrients in deep sediments could be made available by man's activities such as artificial stirring of the sediments or plowing or by natural events such as upheaval.

Under natural conditions, such distinctions may not be quantitatively important to considerations of nutrient cycling during an annual cycle. However, geologic output accumulated as long-term storage in the bottom sediments progressively decreases the volume of the lake ecosystem and is thus an important component in hydrarch succession. Accumulation of such sediments would eventually obliterate the lake ecosystem and provide a substrate for the development of a forest-stream ecosystem within the deciduous forest biome. The development of the forest community could be enhanced because of the presence of the rich pool of nutrients in the sediments accumulated in part by the previous aquatic community. Indeed, the processes of hydrarch succession are readily apparent in the glaciated lake regions of the world; lakes are transient features of the landscape.

The lake ecosystem model, like the terrestrial model, provides a theoretical frame for consideration of ecosystem development. Consider hydrarch succession as it might occur at one point in a lake (Figs. 2 and 3). Initially, ecosystem volume would be delimited by the depth of the lake to the original bottom. Gradually, as geologic output to long-term storage (deep sediments) occurred, the functional volume of the ecosystem would decrease. This trend would increase as the lake became quite shallow and the vegetation dominance shifted from plankton to rooted macrophytes.

As the lake becomes more shallow functional recycling of nutrients stored in the sediments will be facilitated because more frequent, complete circulations of the entire water column are possible, allowing for increased regeneration of nutrients from the mud to the pelagic regions. In addition, rooted macrophytes may utilize nutrients directly from the sediments. In the latter stages of this hydrarch development, the phytoplankton productivity may decrease on a unit area basis, even though it may have reached a maximum on a unit volume basis, because of the limited depth (volume) where photosynthesis can occur.

Eventually, the functional volume of the ecosystem would be at a minimum coincident with the presence of emergent aquatic macrophytes, i.e., the functional top and bottom would be close to the water level. However, as the rooted macrophytes colonize the shallow, nutrient-rich lake sediments, biological productivity may reach its highest level for the site (Westlake, 1963). Thus when the ecosystem volume is minimal biological productivity may well be maximal. It may be that with adequate amounts of available nitrogen and phosphorus, the optimum situation for a plant is to be rooted in a liquid medium, but to get carbon dioxide from the air where differential replacement at the plant surface can be much faster. Plants such as *Typha* spp., *Scirpus* spp., *Carex* spp., and *Phragmites* spp. make the best of both worlds. At this point the ecosystem volume would begin to increase again, primarily as the shoots of these plants grew into the air space above.

In terms of volume, then, a comparison of xerarch and hydrarch succession presents totally different routes to the ultimate developmental stage, with xerarch succession characterized by a sigmoid curve and hydrarch pond succession showing a progressive and relatively more gradual volume decline until there is a shift from a planktonically dominated ecosystem to one dominated by rooted macrophytes, which is then followed by a more sigmoid volume increase comparable to the xerarch ecosystem (Fig. 2).

Within the lake ecosystem, nutrients may occur or exchange between three compartments: available nutrients, organic matter, and primary and secondary minerals, which combined comprise the intrasystem cycle (Fig. 4). Available nutrients are those dissolved in water or on exchange surfaces of pelagic particulate matter or bottom sediments. Nutrients incorporated in living or dead organic matter, both in the pelagic region or in sediments, comprise the organic matter compartment. Nutrients incorporated in rocks, as primary and second-ary minerals in the sediments or suspended in the water, constitute the primary and secondary mineral compartment. The intrasystem cycle of the aquatic system is similar to that of the terrestrial ecosystem, but differs in several important respects.

Aquatic organisms absorb and assimilate available nutrients from the pelagic region as well as from the sediments. Available nutrients are released from organic matter by excretion, exudation, leaching, respira-tion, and decomposition. Primary and secondary minerals may chemi-cally decompose to form available nutrients, or secondary minerals may be reformed from available nutrients. This may be directly related to environmental conditions. For example, ferrous iron and phosphate compounds are soluble under anaerobic and acid conditions, but are transformed to insoluble ferric phosphate by adding oxygen under more alkaline conditions (Einsele, 1936). In such a way, phosphate and iron may cycle between insoluble secondary minerals and soluble available nutrients with the seasonal changes in oxygen content of the hypolimnion in some lakes.

Secondary minerals may be formed from available nutrients by the activity of organisms. Ruttner (1963) reports that *Elodea cana-densis* precipitated $CaCO_3$ at the rate of 0.02 kg/kg fresh weight during 10 hours of sunlight. The very large deposits of marl in some lakes attest to the importance of such mechanisms in the nutrient budgets of lakes. Conversely, some organisms may be able to incorpo-rate directly certain primary or secondary minerals. A number of workers (*see* Hutchinson, 1957: 790) have suggested that diatoms are capable of decomposing aluminosilicate clay minerals to obtain silica, presumably in accordance with their requirements for the element.

Overall it is relatively more difficult to quantitatively measure the nutrient budget for a lake ecosystem since there are a larger number of parameters to be considered. For example, the geologic inputs from the terrestrial watershed must be evaluated, as well as the difficult to measure geologic outputs, including deposition of functionally inactive deep sediments.

As with the forest-stream ecosystem, it may be useful to describe the input-output pathways for an element like calcium through the lake ecosystem model, stressing the similarities and dissimilarities of the two ecosystems.

Calcium may be brought into the lake ecosystem from the water-shed by geologic factors as dissolved inorganic or organic matter in water draining into the lake, as inorganic or organic particulate matter suspended in drainage waters or transported as bed load by the incoming streams, or as inorganic or organic colluvial materials moved downslope and into the lake basin (e.g., by a landslide). Inputs of dissolved calcium would contribute to the available nutrient compartment, while alluvial and colluvial particulate matter would add to the organic matter and primary and secondary minerals compartments, depending upon whether the materials were organic or inorganic.

If there is water drainage from the lake ecosystem, then calcium may be lost in dissolved form from the available nutrient compartment, and to a lesser extent (since the lake is a settling basin) in particulate form from the organic matter and primary and secondary minerals compartments. Thus a relatively large amount of the calcium would normally be deposited as sediments (ultimately), and therefore on a long-term basis represent ecosystem storage in the deep sediments.

Meteorologic inputs of calcium would be identical to those for the forest-lake ecosystem and need no reiteration. As mentioned previously, meteorologic outputs for calcium would be derived from the available nutrient compartment and would be negligible.

The pathways for biological inputs and outputs of calcium are very similar to those in the forest-stream ecosystem. However, the ledger is less likely to balance. For example, insects (e.g., chironomids and ephemeropterans) that live aquatically during the early major part of their life history and later emerge only briefly as flying adults derive all of their structural components directly or indirectly from the aquatic ecosystem. Masses of these organisms are often found dead in the terrestrial ecosystem following an emergence (Kormondy, 1969). Thus significant quantities of calcium (and other nutrients) may be transported by such biological vectors from the lake ecosystem to the surrounding terrestrial ecosystem. Since the land is continually losing certain nutrients that have a sedimentary cycle, such biological returns "uphill" via animal vectors are of interest and ecological significance (*see also* Leopold, 1949; Hutchinson, 1952).

Despite the mind-boggling complexity of ecosystems, man is currently confronted with the necessity of understanding their structure and function. We believe that the general models that we have presented apply to a variety of ecosystems and provide a conceptual device whereby we can simplify the quantification of important functional interactions to facilitate comparison of diverse ecosystems and ultimately to advance our comprehension and management of landscapes. These models allow us to visualize where nutrients are localized within the system and the processes or pathways by which the nutrients are cycled. This knowledge immediately connects phenomena which are often considered separately as biological, geological, pedological, or hydrological factors. The models thus point out the interdependency of isolated bits of information to the functioning whole, and amalgamate the work of different scientific disciplines. They show where it is most advantageous to apply real data for ecosystem analysis. The input and output relationships of the models illuminate the connection between seemingly unrelated ecosystems, and between individual ecosystems and the larger biogeochemical cycles of the earth. The models provide for an integration of classical concepts of biological succession and climax, and at the same time provide a basis for quantitative analysis of degradation resulting from human mismanagement. It is readily possible to conceptualize the ecosystem's reactions and interactions resulting from external or internal perturbation. The models thus provide a simple, but powerful, theoretical framework that incorporates various types of real data in such a way that the interrelationships are readily visible and subject to direct analysis and interpretation.

An Illustration—The Real World

We will use data from a comprehensive study of nutrient-hydrologic interactions in small watershed ecosystems of the Hubbard Brook Experimental Forest to illustrate nutrient cycling in a forest-stream ecosystem of the eastern deciduous forest. Specific data of this sort, in context with theoretical concepts of the ecosystem, are valuable in understanding basic ecological interactions within and between ecosystems, as well as the intricate natural relationships maintaining ecological balance. Moreover, the small watershed approach provides an opportunity to deal with the complex problems of the ecosystem on an experimental basis. In spite of the value of such data, there are few available (Likens *et al.*, 1967). However, currently the ecosystem approach represents one of the major undertakings of the U. S. contribution to the International Biological Program (IBP).

The Hubbard Brook Ecosystem

The Hubbard Brook Experimental Forest is located in the White Mountains of north central New Hampshire. It is maintained and operated by the Forest Service, U. S. Department of Agriculture, as the major research laboratory for management of forested watersheds in New England (U. S. Forest Service, 1964). The Experimental Forest ranges in altitude from 299 to 1,015 m and covers 3,076 ha of rugged terrain (Fig. 5). The forest represents a nearly mature hardwood forest ecosystem, and is characterized by uneven-aged, well-stocked, second growth northern hardwoods with more coniferous species at high elevations and on north-facing slopes. Six small, well-defined watersheds (Watersheds 1 through 6, Fig. 5) have been selected for intensive study. These are all steep (average slope,

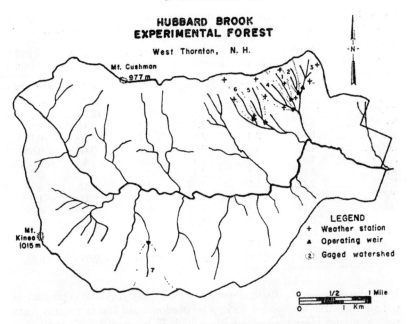

FIGURE 5. Outline map of the Hubbard Brook Experimental Forest showing the gauged watersheds, weather stations, and drainage streams tributary to Hubbard Brook (after Likens *et al.*, 1967).

29 percent), southerly facing watersheds with similar vegetation, till, and bedrock. These watershed ecosystems range in size from 12 to 43 ha, and in altitude from 500 to 800 m.

The major overstory tree species on the experimental watersheds are sugar maple (*Acer saccharum*), beech (*Fagus grandifolia*), yellow birch (*Betula alleghaniensis*), and red spruce (*Picea rubens*), with some white birch (*Betula papyrifera*) and balsam fir (*Abies balsamea*). The forest basal area is about 24 m²/ha. The Experimental Forest was extensively cut about 1919, but no cutting or fire has occurred since. Detailed descriptions of the tree and herbaceous vegetation are given by Bormann *et al.* (1970) and Siccama *et al.* (1970).

Bouldery glacial till covers most of the area and is generically similar to the bedrock lithologies, which are Littleton formation sillimanite-zone gneiss and Kinsman quartz monzonite (Table 1; Johnson *et al.*, 1968). Based upon the amount of available nutrients, weathering rates and vegetation analysis, the area is considered to be relatively oligotrophic (Likens *et al.*, 1967; Fisher *et al.*, 1968; Johnson *et al.*,

Table 1. BULK COMPOSITION OF VARIOUS LITHOLOGY TYPES AND WEATHERING PRODUCTS, HUBBARD BROOK EXPERIMENTAL FOREST (From Johnson et al. 1968)

Products	Littleton formation Unw.	Littleton formation Weath.	Till groundmass	Kinsman formation Unw.	Kinsman formation Weath.	A2 soil horizon
	Percent	Percent	Percent	Percent	Percent	Percent
SiO_2	64.8	70.8	70.5	64.5	69.2	80.8
Al_2O_3	16.2	14.3	14.1	16.1	15.7	9.6
FeO	5.4	1.2	1.5	4.3	0.44	0.64
Fe_2O_3	0.86	3.4	3.4	1.7	3.3	0.16
CaO	1.1	0.87	0.98	2.6	0.17	0.60
Na_2O	1.8	1.5	1.8	2.4	1.6	1.3
K_2O	3.6	3.1	3.3	3.5	5.5	2.9
MgO	2.4	1.1	0.98	1.8	0.42	0.13
TiO_2	1.2	1.0	1.0	1.4	1.1	1.0
MnO	0.14	0.07	0.19	0.12	0.07	0.09
P_2O_5	0.23	0.32	0.12	0.22	0.12	0.02
H_2O^-	0.16	0.49	0.46	0.13	0.31	0.36
H_2O^+	2.1	1.7	1.6	1.3	1.9	1.5
CO_2	<0.05	<0.05	<0.05	<0.05	<0.05	<0.05

1968; Siccama et al., 1970; Bormann et al., 1970). About 6 percent of the surface area is exposed rock. The predominant soil is a sandy loam podzol of the Hermon series with a thick H-layer and a discontinuous but often well-developed A_2 horizon.

The climate is characterized by short, cool summers and long, cold winters. Although the Hubbard Brook ecosystems are only 116 km from the Atlantic Ocean, the climate is predominantly continental, since the atmospheric flow is usually offshore (Likens et al., 1967). The average July air temperature is 19° C and the average January air temperature is –9° C. Significantly, the forest soils usually remain unfrozen during the coldest months because of the thick humus layer and a deep snow cover in winter (Hart et al., 1962). Since there is little soil frost, most water infiltrates the permeable soil at all times and there is very little overland flow (Pierce, 1967). Also, the geologic substrate is thought to be watertight, and losses of water by deep seepage are minimal (Likens et al., 1967). Additional details concerning the topography, climate, geology, and biology of the Hubbard Brook Experimental Forest are given by Likens et al. (1967), Johnson et al. (1968), Bormann et al. (1970), and Siccama et al. (1970).

Methods and Procedures

Our basic procedures have been described extensively (*see*, e.g., Likens et al., 1967) and will only be summarized here.

Precipitation is measured within the experimental watersheds with a network of gauges, approximately one for every 12.9 hectares. The precipitation collectors are located in small cleared areas within the forest. Streamflow is measured continuously at stream-gauging stations, which include a V-notch weir or a combination of V-notch weir and San Dimas flume anchored to the bedrock at the base of each experimental watershed. Continuous measurements of streamflow are made throughout the year.

Weekly samples of precipitation and stream water are obtained from the experimental areas for chemical analysis. Rain and snow are collected both in plastic containers left continuously uncovered and in containers uncovered only during periods of rain or snow. One-liter samples of stream water for chemical analysis are collected in clean polyethylene bottles approximately 10 m above each weir. Long-term average concentrations (mg/liter) determined arithmetically from weekly samples may be misleading, particularly for precipitation, because small amounts of water with high salt content will contribute disproportionately to the final value. Therefore chemical concentrations characterizing a period of time are reported as averages weighted by volume of water.

Hydrologic Inputs and Outputs

The U. S. Forest Service has collected hydrologic information for the Hubbard Brook Experimental Forest since 1955. Annual precipitation averages about 123 cm (Table 2), of which about one-third to one-fourth is snow. On the average, some 59 percent of this water is lost as runoff, and the remaining 41 percent is lost as evaporation and transpiration. The monthly precipitation input as rain and snow is

Table 2. AVERAGE HYDROLOGIC BUDGETS FOR WATERSHEDS 1-6, HUBBARD BROOK EXPERIMENTAL FOREST (After Likens *et al.*, 1971)

Water-year*	Precipitation	Runoff		Evapotranspiration†	
	cm	*cm*	*Percent*	*cm*	*Percent*
1963-64	117.1	67.7	57.8	49.4	42.2
1964-65	94.9	48.8	51.4	46.1	48.6
1965-66	124.5	72.7	58.4	51.8	41.6
1966-67	132.5	80.6	60.8	51.9	39.2
1967-68	141.8	91.2	64.3	50.6	35.7
1968-69	128.3	86.9	67.7	41.4	32.3
1963-69‡ (mean)	123.2	74.7	60.6	48.5	39.4
1955-69§ (mean)	123.1	73.0	59.3	50.1	40.7

* Water-year is 1 June to 31 May.

† Precipitation minus runoff.

‡ Runoff from Watershed 2 not included 1965-69.

§ The number of watersheds used to calculate the mean for each year varied during the period.

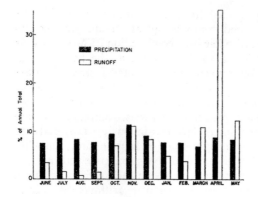

FIGURE 6. Average annual distribution of precipitation and runoff, 1955-1969, for the Hubbard Brook Experimental Forest (after Likens et al., 1971).

relatively constant throughout the year (Fig. 6). On the average, more precipitation falls in November (13.9 cm), and somewhat less falls in March (8.3 cm), but no average monthly value deviates widely from the long-term monthly average of 10.3 cm. Snow accumulates in the area during three to four months of winter (usually early December to early April), reaching a peak accumulation of 90 to 150 cm in March (Hornbeck and Pierce, 1969; Pierce et al., 1970).

Runoff varies markedly with season (Fig. 6). Sixty-eight percent of annual runoff occurs during the snowmelt period of March, April, and May, with more than 35 percent in April alone. In contrast, runoff during the active growing season is small, with only 0.5 cm, or 0.7 percent, of the annual runoff occurring in August. The small bouldery drainage streams sometimes have intermittent flow during the summer or early autumn. During this period the percentage of water lost by transpiration from leaf surfaces may be relatively large. The importance of this factor is often vividly demonstrated in the autumn when, as the deciduous leaves begin to fall, streamflow increases without any input of water from precipitation.

Since the geologic substrate is watertight, quantitative measures of annual evapotranspiration can be obtained by difference. A range of values for precipitation, runoff, and evapotranspiration, including an exceptionally dry year (1964-65) and a wet year (1967-68), is given in Table 2. During the drought year not only was the precipitation 28.2 cm less than the 1955-1969 average, but the percentages of streamflow and evapotranspiration were altered. Overall evapotranspirational losses were more constant than runoff losses or precipitation inputs during 1963-1969 (Table 2).

Chemical Inputs

Chemical input is calculated by multiplying the chemical concentration (mg/liter) times the volume (liters) of precipitation. These chemical input data represent bulk precipitation, i.e., a mixture of rain or snow and dry fallout (Whitehead and Feth, 1964). We have been concerned with the contributions from dry fallout since the beginning of the study.

Wind-borne dust from dirt roads and agricultural activity may constitute a major input of Ca^{++}, Mg^{++}, K^+, and Na^+ into watersheds (Tamm and Troedsson, 1955; Holstener-Jørgenson, 1960; Gambell and Fisher, 1966). It seems unlikely that locally derived soil dust contributes significantly to the ionic concentration of precipitation in the Hubbard Brook ecosystem. These watersheds are situated within the extensive experimental forest, which is part of the much larger White Mountain National Forest. There are relatively few roads within the forest and traffic is infrequent. Probably less than one percent of the surrounding country is under cultivation.

Juang and Johnson (1967), based on the assumption that in one year the input of chloride should balance output, suggested that dry fallout was a source of chloride in the Hubbard Brook ecosystem. Subsequently, from comparisons between precipitation collectors that were continuously open and those that opened only during periods of rain or snow, it has become clear that the bulk of chemical input to the ecosystem comes directly in rain and snow (Likens *et al.*, 1970; Johnson *et al.*, 1969). Our attempts to quantify the much smaller contributions from dry fallout have been thus far inconclusive.

Sulfate and hydrogen ions are the most abundant constituents (in terms of chemical equivalents) of precipitation falling on the experimental watersheds. The pH of rain and snow samples is frequently less than 4.0. Nitrate is next in abundance, and significant amounts of ammonium, chloride, sodium, and calcium are present. Lesser amounts of magnesium and potassium are also found (Table 3). A study undertaken in 1964-65 indicated that there was no significant difference in cation precipitation chemistry throughout the experimental watersheds (Likens *et al.*, 1967).

The weighted concentration of the various cations and anions in precipitation may range over almost three orders of magnitude on a weekly basis. Since precipitation samples are obtained during weekly intervals, it is difficult to assign the origin of the chemical impurities to individual air masses or storms. Recurring seasonal changes are not entirely obvious, although on the average relatively lower values usually occur in winter (Fig. 7; Likens *et al.*, 1967; Fisher *et al.*, 1968; Likens *et al.*, 1971). The cations usually show maximum concentrations in May

Table 3. Weighted average concentrations of various dissolved substances in bulk precipitation and stream water for undisturbed Watersheds 1-6, Hubbard Brook Experimental Forest, 1963-1969

	Precipitation	Stream water
	mg/liter	mg/liter
Calcium	0.21	1.58
Magnesium	0.06	0.39
Potassium	0.09	0.23
Sodium	0.12	0.92
Aluminum	*	0.24
Ammonium	0.22	0.05
Sulfate	3.1	6.4
Nitrate	1.31	1.14
Chloride	0.42	0.64
Bicarbonate	*	1.9†
Dissolved silica	*	4.61

* Not determined, but very low.
† Watershed 4 only.

or June with secondary peaks in the autumn (Fig. 7). The year-to-year variation in precipitation chemistry is relatively small, although the concentrations of calcium, magnesium, potassium, and sodium in precipitation during the drought year, 1964-1965, were appreciably higher than the mean values for the period 1963-1969 (Likens et al., 1971).

Nitrate and sulfate values are currently higher (Table 3) than reported for the area in 1955-1956 (Junge, 1958; Junge and Werby, 1958). These increases may reflect some measure of increased air pollution.

Chemical Outputs

Dissolved substances: One of our first considerations was to determine the sampling intensity necessary to characterize accurately the stream-water chemistry. Initially we collected samples every hour during a 24-hour period, during rainstorms and at different elevations within the watersheds. Also we installed a continuously recording electrical conductivity unit to monitor the stream-water chemistry. However, after the first year, 1963-64, it became apparent that even though the stream-flow discharge ranged over four orders of magnitude during an annual cycle, the weekly chemical concentrations remained relatively constant (Likens et al., 1967). This is particularly true for magnesium, sulfate, chloride, and calcium concentrations. Sodium and silica concentrations may be diluted up to three-fold, while aluminum,

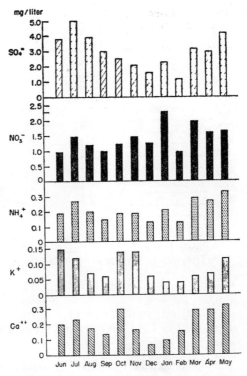

FIGURE 7. Weighted monthly average nutrient concentrations in precipitation during 1963-1969 for Watershed 2 of the Hubbard Brook Experimental Forest. Values for NH_4^+, NO_3^- and $SO_4^=$ are for 1965-1969 only.

nitrate, hydrogen ion, and potassium concentrations are increased with increased discharge (Table 4; Johnson *et al.*, 1969). Biologic activity appreciably reduces the concentration of nitrate and potassium in stream water during the summer. Of the major alkalies and alkaline earths, potassium is by far the most sensitive index of annual biologic activity. Its concentration in stream water sharply decreases during periods of plant growth and increases during plant dormancy. Johnson *et al.* (1969) have developed a working model to describe the seasonal variation in stream-water chemistry as a function of stream discharge.

Year-to-year variations in stream-water chemistry also are relatively small in comparison to yearly variations in the hydrologic cycle (Likens *et al.*, 1971). Sulfate and calcium are the dominant ions in stream water at Hubbard Brook (Table 3). Bicarbonate is a minor constituent of the acidic drainage waters from the experimental watershed ecosystems. The weighted average pH of the stream water is usually between 5.0 and 6.0, and is variable throughout the year.

Table 4. General behavior of major ions in stream water from Watersheds 1-6, Hubbard Brook Experimental Forest (Modified from Johnson et. al., 1969)

	Chemically buffered (limited extremes)	Dilution with stream discharge	Concentration with stream discharge
Na+	*	++	—
SiO₂-aq.	*	++	—
Mg++	‡	+	—
SO₄⁻	‡	+	—
Cl⁻	‡	—	—
Ca++	†	+	+
Al+++	*	—	++
H+	*	—	+
NO₃⁻	*	—	++
K+	*	—	+

* Poorly buffered   † Medium buffered   ‡ Highly buffered
\+ Irregular occurrence   ++ Consistent occurrence   — Not manifested

Electrical conductivity of the stream water averages about 20 μmhos/cm² at 25° C and changes very little either on an hourly, daily, or seasonal basis.

The maintenance of relatively stable stream-water chemistry is apparently due to several factors, including: (1) the almost complete infiltration of water into the soil, rather than runoff as overland flow (Pierce, 1966); (2) the relative absence of soil frost during the winter (Hart *et al.*, 1962); (3) the major loss of the incoming precipitation (some 60 percent, Table 2) as stream discharge; and (4) the abundance of exchangeable cations in relation to amounts lost in transient waters. Thus, because most of the incoming precipitation percolates through the soil, the various chemical equilibria are achieved rapidly through geologic and biologic processes in the soil.

Since the chemical concentrations are relatively constant in stream water, the gross export of dissolved substances is directly related to the amount of stream discharge (Fig. 8). Hence the output of cations and anions is low during the drought years (1963-65) and high during the wet years (e.g., 1967-68). Seasonally the output is low during summer months when runoff is small and high during periods of peak runoff, e.g., April and November (Likens *et al.*, 1967; Fisher *et al.*, 1968). Other biologic and geologic factors may alter this latter relationship slightly (Johnson *et al.*, 1969).

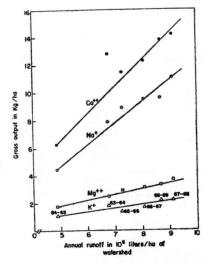

FIGURE 8. Relationship between the gross output of the major cations and annual water discharge during 1963-1969 for the Hubbard Brook watershed ecosystems. The year of occurrence is shown only for the potassium data, but applies to the other ions as well (modified from Likens *et al.*, 1971).

PARTICULATE MATTER: Organic and inorganic particulate matter is also lost from the watershed ecosystems through the processes of erosion and transport by stream water. We estimate the particulate matter output from routine measurements of the suspended and bed load matter that settles in the stilling basin upstream from the weir, the suspended or floating materials larger than 1 mm (collected with a 1-mm mesh net below the V-notch at the weir), and the suspended and floating matter <1 mm but >0.45 μ, collected from subsamples of the water passed through the 1-mm mesh net.

A total of about 25 kg/ha-yr of particulate matter is lost from the undisturbed watershed ecosystems. In contrast the total losses of dissolved substances (including organic matter) average about 140 kg/ha-yr (Bormann, Likens, and Eaton, 1969). Moreover, considering the output of the major elements except carbon, oxygen, and hydrogen, about 90 percent is exported as dissolved substance and 10 percent in organic and inorganic particulate matter (Table 5). In each individual case, except iron, the great bulk is exported as dissolved substances. In contrast to the dissolved substances, particulate matter concentrations were directly related to stream discharge (Fig. 9). The losses of particulate matter showed seasonal variations, with the maximum export during spring snow melt and subsequent high runoff. However, hydrologic discharge is less effective in removing particulate matter during the dormant season than during the growing season. A single, large

Table 5. Average annual gross losses of some elements as dissolved and particulate substances from undisturbed Watershed 6, 1965-67 (Modified from Bormann, Likens, and Eaton, 1969)

Element	Particulate	Dissolved	Total	Particulate	Dissolved
	kg/ha	kg/ha	kg/ha	Percent	Percent
Aluminum ..	1.00	2.60	3.60	1.55	4.03
Calcium	0.25	10.55	10.80	0.39	16.35
Iron	0.50	*	0.50	0.78	*
Magnesium ..	0.17	2.80	2.97	0.26	4.34
Nitrogen	0.12	1.90	2.02	0.19	2.95
Potassium	0.33	1.50	1.83	0.51	2.33
Sodium	0.19	6.55	6.74	0.29	10.15
Silicon	3.69	15.92	19.61	5.72	24.68
Sulphur	0.02	16.42	16.44	0.03	25.45
Total	6.27	58.24	64.51	9.72	90.28

* Not measured, but very small.

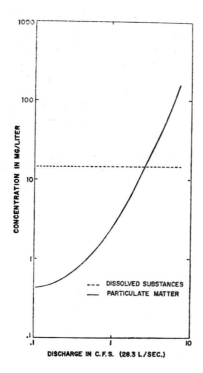

Figure 9. General relationships between the concentration of dissolved substances and particulate matter and stream discharge in a northern hardwood forest ecosystem (after Bormann, Likens, and Eaton, 1969).

Table 6. SODIUM BUDGET FOR FIVE UNDISTURBED WATERSHEDS, HUBBARD BROOK
EXPERIMENTAL FOREST, 1968-69.

Watershed	Input	Output	Net loss
	kg/ha	kg/ha	kg/ha
W1	1.0	6.8	5.8
W3	1.1	7.9	6.8
W4	1.1	9.7	8.6
W5	1.1	6.7	5.6
W6	1.1	7.0	5.9
Mean ± Standard Error	1.1±0.02	7.6±0.56	6.5±0.55

autumnal storm flushed 54 percent of the total particulate matter output during a 2-year period (Bormann, Likens, and Eaton, 1969).

Nutrient Budgets for Dissolved Substances

Quantitative nutrient budgets for the Hubbard Brook watershed ecosystems are determined from the difference between the meteorologic input and the geologic output (Fig. 1). Input is calculated as the product of chemical concentration (mg/liter) times the volume (liters) of precipitation, and the output is the product of the volume (liters) of water draining the watershed ecosystem times its chemical concentration (mg/liter). These are nutrient budgets for dissolved substances, and it should be remembered that there is an additional small loss of potential nutrients in particulate matter, as indicated above. Since measurements are made simultaneously on several adjacent, similar watersheds, statistical estimates of variation in chemical inputs and outputs can be determined (e.g., Table 6). These watershed ecosystems thus may be considered as replicated, experimental units at the ecosystem level, and are amenable to experimental manipulation.

Although calculations are made on a weekly basis for each watershed, results are summed by water-year (1 June to 31 May). Nutrient budgets vary seasonally primarily because of variations in the hydrologic inputs and outputs (Fig. 10). Since nutrient budgets are strongly geared to the hydrologic cycle, budgets for individual years differ appreciably (Table 7). These long-term, replicated studies show that there is a significant annual net loss of dissolved silica, calcium, sulfur, sodium, magnesium, and aluminum and a net gain of nitrate nitrogen and ammonium nitrogen (gaseous exchanges are not considered) for these watershed ecosystems. Crisp (1966) also observed that the outputs of nitrogenous substances dissolved in stream water were less than the inputs in precipitation for an area dominated by blanket bog vegetation in Great Britain. However, enough nitrogen was lost as eroding

Table 7. Average chemical input and output for undisturbed, forested watershed-ecosystems W1-W6, Hubbard Brook Experimental Forest, 1963-1969. (Year [e.g. '5 = 1964-1965] and watershed number [e.g., W2] indicated for minimum and maximum values; output for W2 after treatment not included)

Element	Input			Output			Mean values		
	Water-shed years*	Range		Water-shed years*	Range		Input	Output	Net loss or gain
		Minimum	Maximum		Minimum	Maximum			
		kg/ha-yr	kg/ha-yr		kg/ha-yr	kg/ha-yr		kg/ha-yr	
SiO_2-Si	0	....	....	10	9.7 ('5 W2)	22.3 ('8 W4)	†	16.4	—16.4
Ca^{++}	30	1.6 ('9 W2)	3.0 ('8 W6)	32	4.8 ('5 W6)	17.3 ('9 W4)	2.6	11.7	— 9.1
SO_4-S	15	9.7 ('5 W2)	16.0 ('8 W6)	10	9.8 ('5 W6)	19.6 ('8 W4)	12.7	16.2	— 3.5
Na^+	30	1.0 ('9 W2)	2.3 ('5 W6)	32	3.5 ('5 W1)	11.0 ('8 W4)	1.5	6.8	— 5.3
Mg^{++}	30	0.32 ('9 W2)	1.2 ('5 W6)	32	1.6 ('5 W3)	3.7 ('8 W1)	0.7	2.8	— 2.1
Al^{+++}	0	....	....	10	0.9 ('6 W4)	3.2 ('9 W6)	†	1.8	— 1.8
K^+	30	0.57 ('7 W3)	2.4 ('4 W4)	32	0.76 ('5 W3)	2.7 ('9 W1)	1.1	1.7	— 0.6
NO_3-N	15	1.5 ('5 W2)	5.2 ('8 W6)	10	1.1 ('5 W6)	2.9 ('8 W4)	3.7	2.0	+ 1.7
NH_4-N	15	1.6 ('5 W2)	2.6 ('8 W6)	10	0.08 ('9 W6)	0.7 ('6 W6)	2.1	0.3	+ 1.8
Cl^-	12	2.6 ('6 W2)	6.9 ('7 W6)	8	4.2 ('6 W4)	5.3 ('8 W4)	5.2	4.9	+ 0.3
HCO_3-C	0	....	....	4‡	2.4 ('6 W4)	3.3 ('7 W4)	†	2.9	— 2.9

* Number of watersheds times years of data.
† Not measured but very small.
‡ Watershed 4 only.

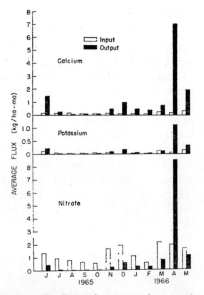

FIGURE 10. Average monthly flux of calcium, potassium, and nitrate in the Hubbard Brook watershed ecosystems during 1968-69. Annual totals showed a significant net loss of calcium (–12.2 kg/ha), a small net loss for potassium (–1.6 kg/ha) and a net gain for nitrate (+3.1 kg/ha).

peat to produce a net loss for the system. Budgets for potassium and chloride are nearly balanced at Hubbard Brook on a long-term basis. There was a net gain of potassium during 1963-65, and net losses during 1965-69 (Likens et al., 1971). A small, but consistent, net loss of bicarbonate occurs from Watershed 4 only. These budgets represent good averages for these undisturbed watershed ecosystems, since the major cation budgets for the period 1963-69 are based upon 30 watershed-years of data (number of watersheds times years of data available).

These long-term records accurately show the biogeochemical interrelationships of these smaller ecosystems to the larger biospheric cycles. Large differences in input and output of nutrients occur from year to year, depending primarily upon variations in the hydrologic cycle and to a lesser extent on ionic concentrations in precipitation and stream water. These results also demonstrate the need for long-term studies to provide reliable biogeochemical data for landscape planning. For example, potassium input exceeded the gross output during 1963-64 and 1964-65, but during the last four years, gross output has exceeded the input (Likens et al., 1971). The amount of precipitation and runoff for the area during 1963-64 and 1964-65 was appreciably less than that during the other four years of the study, or than the 14-year average (Table 1). Firm conclusions, or land managerial applications, based on the first two years of this study alone would be quite misleading, if not ecologically harmful.

Weathering Rates

Net losses of dissolved materials originate from within the ecosystem. If the organic components of the ecosystem are near equilibrium, the net loss of dissolved nutrients is a direct estimate of the rate of biogeochemical weathering of primary and secondary minerals within the ecosystem (Bormann and Likens, 1967). Based on the species composition and structure of the vegetation (Bormann et al., 1970; Siccama et al., 1970), our study area is not quite a fully mature ecosystem. Thus weathering rates based upon nutrient budget calculations may be slightly underestimated, since budget calculations do not account for net storage in accumulating biomass. However, based upon more conservative sodium budgets, Johnson et al. (1968) have estimated that 800 ± 70 kg/ha of till and bedrock undergo chemical decomposition in the Hubbard Brook ecosystem each year. At this rate some 11 x 10⁶ kg/ha of bedrock and till would have been completely weathered since the last glaciation, some 14,000 years ago.

Nutrient Cycling in Undisturbed Forested Ecosystems

Data on inputs, outputs, and weathering may be used, in context with our model for the terrestrial watershed ecosystem (Fig. 1), to describe the various parameters of the nutrient cycle for the Hubbard Brook ecosystems. Data on organic debris, leaching, and decomposition are from studies by J. R. Gosz and J. S. Eaton; estimates of tree biomass and productivity are provided by R. H. Whittaker. At present, we estimate the quantities of nutrients in the primary and secondary minerals, available nutrients, and organic debris compartments from studies by Lunt (1932) on similar areas in New Hampshire. These estimates may be refined as additional data on compartment sizes and exchange rates are obtained from studies currently underway at Hubbard Brook. We will use data for calcium to illustrate these relationships at Hubbard Brook (Fig. 11).

Some 570 kg calcium/ha are held by the vegetation and 1,740 kg calcium/ha in organic debris; some 690 kg/ha are available in soil water and on exchange surfaces; and the soil and rock minerals contain about 28,550 kg calcium/ha. Annual uptake by the vegetation slightly exceeds 49 kg/ha, while about 49 kg/ha are released by decomposition and leaching. Thus about 1/14 of the available calcium pool is cycled through the vegetation each year. The average annual input (1963-69) of calcium is 2.6 kg/ha and the output is 12.0 kg/ha, of which 11.7 kg/ha is lost in dissolved form in the stream water. These data, applied to the model, show that weathering generates on the average some 9.1 kg calcium/ha-yr (Fig. 11). They also indicate that the ecosystem is efficient in retaining and circulating calcium. The annual

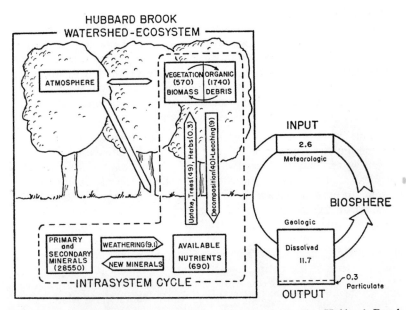

FIGURE 11. Major parameters of the calcium cycle in the Hubbard Brook
watershed ecosystems. All data in kg/ha and kg/ha-yr. Data on
organic debris, leaching, and decomposition from Gosz and Eaton
(*unpublished*) ; vegetation biomass and uptake from Whittaker (*unpublished*) ; primary and secondary minerals, available nutrients,
and organic debris (all to a depth of 61 cm) from Lunt (1932).

net loss (9.1 kg/ha) represents only about 1.3 percent of the total
available calcium (690 kg/ha) in the ecosystem, and about 18 percent of
the amount circulated annually by the vegetation through uptake and
release.

These quantified nutrient relationships (Fig. 11) provide valuable information for conservers, users, and managers of forest ecosystems. The nutrient capital in the various compartments can be as-
sessed, as well as the exchange rates. With such information, it is
possible to make intelligent decisions concerning the wise use and
ecological conservation of such areas. Without this knowledge, such
decisions are exceedingly difficult and often wrong.

Experimental Manipulation of an Ecosystem—
Deforestation

All of the vegetation on one of the experimental watersheds
(Watershed 2) was cut during November and December of 1965,

and vegetation regrowth was inhibited for three summers by the periodic application of herbicides (Fig. 12). Trees, saplings, and shrubs were dropped in place and limbed so that no slash was more than 1.5 m above ground. No timber or other vegetation were removed and no roads or skid trails were made in the watershed. Measurements of precipitation and stream-water quantity and quality were continued after treatment, and similar measurements on adjacent, undisturbed watersheds provided comparative information.

This represented an experimental manipulation at the ecosystem level designed to determine the effects on the quantity and quality of stream water output from the watershed and on the fundamental chemical relationships within the intrasystem cycle of the forest ecosystem. In effect this experiment was designed to test the homeostatic potential of the ecosystem after elimination of an important function, the uptake transfer between the available nutrient compartment and the living organic matter compartment (Figs. 1 and 10). The small watershed approach allows an evaluation of such an experimental manipulation in terms of the entire ecosystem and its interrelationships rather than isolated parts. In this way various land-management practices may be tested and evaluated in terms of the functioning whole.

Hydrologic Effects

Deforestation of Watershed 2 produced a significant effect on the distribution of water loss from the ecosystem (Hornbeck *et al.*, 1970; Pierce *et al.*, 1970). Annual runoff of water exceeded the expected value, had the watershed not been cut, by 35 cm, or 40 percent, during the first year; 27 cm, or 28 percent, during the second year; and 24 cm, or 26 percent, during the third water-year after deforestation. The largest increase in water discharge, relative to an undisturbed condition, occurred during June through September, when stream runoff was 414 percent (first year) and 380 percent (second year) greater than expected. The increased streamflow during the summer is directly attributable to the removal of transpiring surfaces.

In addition to the increased annual runoff, the snow melted earlier and more rapidly on the cleared watershed. Hornbeck and Pierce (1969) found that the cumulative streamflow was advanced by as much as 17 days or an average of 4-8 days during the period of major snowmelt contribution to runoff. They also observed that the instantaneous peak flows during the early part of the snowmelt period were increased significantly (P < .05), but the total volume of streamflow during the period changed very little. Moreover, it was estimated that forest clearing of only 25 percent of a large south-facing watershed probably would not influence downstream flooding. However, flooding is an

FIGURE 12.   Deforested Watershed 2 of the Hubbard Brook Experimental Forest. This picture was taken in the autumn of 1968.

important consideration in large-scale forest manipulations, particularly in an area such as this where heavy spring rains combined with melt water from a deep snow pack frequently produce flooding. It is obvious that a planned or managed runoff distribution is more economical and beneficial for man's utilization, and forest manipulations can directly affect this distribution.

Chemical Effects

INPUTS: As far as we know normal meteorologic precipitation inputs were not significantly affected by the forest clearing treatment. However, the extraneous chemical inputs from the herbicide applications were significant and were accounted for in the nutrient budgets. Assuming complete decomposition of the herbicides some 0.04 kg Ca^{++}/ha, 0.09 kg Mg^{++}/ha, 0.01 kg K^+/ha, 0.11 kg Na^+/ha, 10.6 kg NO_3^-/ha, and 3.7 kg Cl^-/ha were added to the deforested watershed in herbicides during the first two water-years, 1966-68 (Likens et al., 1970).

It has been suggested (Cooper, 1969) that felling the vegetation without timber removal contributes a large and unnatural input of

Table 8. COMPARATIVE NET LOSSES OR GAINS FOR UNDISTURBED (W6) AND DEFORESTED (W2) WATERSHEDS, HUBBARD BROOK EXPERIMENTAL FOREST

Element	1966-67		1967-68		1968-69	
	W6	W2	W6	W2	W6	W2
	kg/ha	kg/ha	kg/ha	kg/ha	kg/ha	kg/ha
Ca	—8.3	—75.0	—9.2	—90.4	—10.0	—68.2
Mg	—2.5	—15.7	—2.6	—17.9	— 2.7	—13.2
K	—1.1	—22.5	—1.6	—35.8	— 1.7	—32.8
Na	—5.5	—16.8	—7.0	—17.3	— 5.8	—12.2
Al	—1.4	—16.9	—3.1	—24	— 3.2	—21
NH_4-N	+1.6	+ 1.1	+2.4	+ 1.9	+ 2.4	+ 1.9
NO_3-N	+3.3	—97.2	+2.4	—142	+ 1.0	—103
SO_4-S	—2.7	— 1.7	—3.3	0	— 6.3	— 6.6
Cl	+2.3	— 1.1	—0.1	— 3.7	+ 1.3	— 1.1
HCO_3-C	—0.4	— 0.2	—0.5	0	—	0
SiO_2-Si	—17	—31	—17	—32	—14	—28

nutrients to the ecosystem, available upon decomposition. However, this was not the case at Hubbard Brook, since the cutting occurred after the nutrient-rich leaves had fallen naturally, and the boles (wood) of the trees decompose very slowly and are characterized by low nutrient content (Likens and Bormann, 1971).

OUTPUTS: Deforestation resulted in large increases in stream-water concentrations for all major ions except NH_4^+ and $SO_4^=$ (Fig. 13; Likens et al., 1970). However, the increases did not occur until about five months after the forest was cleared. Nitrate concentrations increased spectacularly, from a weighted concentration of 0.9 mg/1 prior to cutting to 53 mg/1 two years later. Measured concentrations of nitrate soared to 82 mg/1 in October, 1967 (Fig. 13).

The deforestation treatment resulted in an alteration of the basic nitrogen cycle for the ecosystem. Whereas nitrate-nitrogen and ammonium-nitrogen are normally conserved by the undisturbed forest ecosystem, in the deforested condition nitrate is flushed from the system in large amounts in drainage water (Table 8). The mobilization of nitrate from decomposing organic matter, presumably from increased microbial nitrification, quantitatively accounted for the net increase in total export of cations and anions from the deforested ecosystem (Likens et al., 1969).

To understand this change in the nitrogen cycle, we should briefly review what we know of the cycle in the undisturbed forest. Since yearly input of nitrate-nitrogen in precipitation exceeds the losses in stream water, the presence of nitrate in stream water provides no conclusive evidence for the occurrence of nitrification in these acid

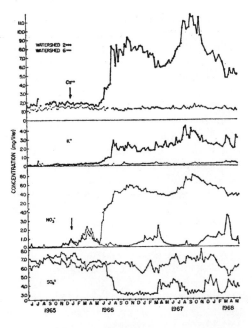

FIGURE 13. Measured stream-water concentrations for calcium, potassium, nitrate, and sulfate in Watersheds 2 (deforested) and 6. Note the change in scale for the nitrate concentration. The arrow indicates the completion of cutting in Watershed 2. (Modified from Likens *et al.*, 1970.)

soils. Many microorganisms decompose nitrogenous organic materials to ammonia. The ammonium ions then may be taken up and assimilated by green plants, or they may be oxidized by bacteria to nitrite and nitrate. The low levels of ammonium and nitrate in the drainage waters of the undisturbed watershed (Table 3) may attest to the efficiency of the oxidation of ammonia to nitrate, and to the efficient uptake of nitrate by the vegetation. However, Nye and Greenland (1960) and others state that growing, acidifying vegetation in acid podzol soils represses nitrification. Thus we believe that the bulk of the ammonia generated from decomposition in the undisturbed watershed ecosystem is used directly by green plants and little of it is converted to nitrate. However, in the absence of green plants, the microflora of the deforested watershed apparently oxidize ammonia to nitrate and the very mobile nitrate ions are rapidly flushed from the watershed ecosystem. High concentrations of nitrate have occurred in stream water for three years following deforestation.

The herbicide apparently reinforced the already well-established trend of nitrate loss established after the vegetation was cut. This probably acted through the destruction of the remaining vegetation (including roots) and the prevention of vegetation regrowth. Exchanges of gaseous nitrogen were not measured, but it is probable that both microbial fixation and denitrification are minimal in these acid forest soils (C. O. Tamm, personal communication). However, Alexander (1967) points out that significant chemical volatilization of nitrogen may occur when ammonium oxidation occurs at a pH lower than 5.0 to 5.5.

Hydrogen ions are produced by nitrification and some of these replace metallic ions on exchange surfaces in the soil. Stream water from the undisturbed watersheds can be characterized as a very dilute solution of sulfuric acid (pH about 5.1 for W2); however, after deforestation the stream water from Watershed 2 became a relatively stronger nitric acid solution (pH 4.3), considerably enriched in metallic ions and dissolved silica (Likens *et al.*, 1970).

Average net losses of nitrate-nitrogen were 97 kg/ha in 1966-67, 142 kg/ha in 1967-68, and 103 kg/ha in 1968-69, or a total of 342 kg/ha during the three years following deforestation (Table 8). If nitrogen in precipitation were the only source for the ecosystem and if it were all retained each year, it would take about 43 years to replace the nitrogen mobilized and lost in these years from the deforested watershed ecosystem (Input $= NH_4 -N + NO_3 -N +$ organic $-N$ $=$ about 8 kg/ha-yr; Table 7). Preliminary estimates indicate that the content of organic-nitrogen and ammonium-nitrogen are about equal for the Hubbard Brook area.

These losses of nitrogen are very large in comparison with some other data that are available. Sylvester (1961) showed an average loss of 0.2 m tons/km²-yr in rivers draining forested areas in Oregon. Sawyer (1947) estimated that about 0.8 m tons of N/km²-yr were lost in agricultural drainage around Madison, Wisconsin. Neil, Johnson, and Owen (1967) found in the Toronto area the total nitrogen loss in rural streams ranged between 0.3 and 0.8 m tons/km²-yr and in urban streams, receiving effluent from sewage treatment plants, averaged 6.0 m tons/km²-yr. At Hubbard Brook the net loss of nitrogen in nitrate alone averaged 11.4 m tons/km²-yr from the deforested area (Table 8).

Net losses from the deforested watershed during 1967-68 (the second year) were greater than the undisturbed condition by 9.8-fold for Ca^{++}, 6.9-fold for Mg^{++}, 22.4-fold for K^+, 2.5-fold for Na^+, 7.7-fold for Al^{+++}, 37-fold for Cl^-, and 1.9-fold for dissolved Si (Table 8). The cations were mobilized as hydrogen ions replaced them on various exchange complexes in the soil and as organic and inorganic materials were decomposed. It was estimated that chemical decomposition of inorganic materials was increased about three-fold; thus the bulk of the nutrient export from the deforested watershed originated from the organic compartment of the watershed (Likens *et al.*, 1970).

The average net export of dissolved inorganic substances from the deforested watershed during 1966-1969 increased about 13-fold relative to the undisturbed ecosystems (Likens *et al.*, 1970; or as calculated from Table 8). This increased net export occurred then because the stream-water concentrations were vastly increased, primarily as a direct result of the increased nitrification, and to a much lesser extent because the streamflow was increased.

In contrast to all of the other major ions, the sulfate concentration in stream water decreased after deforestation. The ammonium concentration also increased, but no more than on adjacent, undisturbed watersheds. The concentration decreased by about 45 percent during the first year after cutting. Net losses of SO_4-S were about 40 percent lower the first year, 100 percent lower the second year, and then returned to a relatively high value the third year after deforestation (Table 8). This interesting and unusual (in relation to the other ions) response during the first two years is discussed in some detail by Likens *et al.* (1970). Briefly, most of this change can be explained by dilution resulting from an increase in stream discharge; however, the normal relatively small release of $SO_4^=$ from the undisturbed ecosystem also became negligible after deforestation. This may have resulted from either decreased oxidation of various sulfur compounds within the ecosystem or increased sulfate reduction and the increased production of sulfides (e.g., H_2S) which are not measured. It is very possible that both of these mechanisms, plus other unknown factors such as nitrate or herbicide toxicity, were affecting the microbial production of sulfate in the deforested ecosystem. The net loss of sulfate during 1968-69 was unusually large from the deforested watershed as well as from the undisturbed watershed. This resulted primarily because the sulfate input was about 28 percent less than the average for the preceding three years, whereas the gross output of sulfate was approximately the same. These results pose some important problems that will require further research.

Phosphorus is typically very low in drainage waters at Hubbard Brook. Average concentrations are about 1 to 2 μg/1 (J. E. Hobbie, personal communication). Although the phosphorus is held very tenaciously by the forest soil, our preliminary estimates indicate that dissolved and particulate phosphorus losses were significantly increased after deforestation.

Total gross export of dissolved substances, exclusive of dissolved organic matter, averaged about 82 metric tons/ha-yr from the deforested watershed and 13 m tons/ha-yr from an undisturbed watershed (W6) during the three years after deforestation (Table 9).

Table 9. COMPARATIVE GROSS LOSSES OF DISSOLVED SUBSTANCES IN STREAM WATER FROM UNDISTURBED (W6) AND DEFORESTED (W2) WATERSHEDS, HUBBARD BROOK EXPERIMENTAL FOREST

Element	Metric tons/km²/yr					
	1966-67		1967-68		1968-69	
	W2	W6	W2	W6	W2	W6
Ca⁺⁺	7.7	1.1	9.3	1.2	7.0	1.2
K⁺	2.3	0.2	3.6	0.2	3.3	0.2
Al⁺⁺⁺	1.8	0.3	2.5	0.3	2.1	0.3
Mg⁺⁺	1.6	0.3	1.9	0.3	1.3	0.3
Na⁺	1.8	0.7	1.9	0.9	1.3	0.7
NH₄⁺	0.09	0.04	0.07	0.02	0.06	0.01
NO₃⁻	46.0	0.6	65.2	1.2	46.9	1.2
SO₄⁻	4.6	5.1	4.5	5.8	5.0	5.1
HCO₃⁻	0.1	0.2	0	0.3	0	0.1
Cl⁻	1.1	0.5	0.9	0.5	0.7	0.5
SiO₂–aq.	6.7	3.7	7.0	3.6	5.9	2.9
TOTAL	73.8	12.7	96.9	14.3	73.6	12.5
THREE-YEAR AVERAGE					81.4	13.2

Coupled with this increased export of dissolved substances was a nine-fold increase in the output of inorganic and organic particulate matter from the deforested ecosystem. Increased erosion of particulate materials occurred as the biotic mechanisms, such as fallen leaf cover and the binding action of rootlets, which normally minimize erosion, were removed or made less effective.

Unexpectedly the stream water from the deforested watershed appeared to be just as clear and potable as that from adjacent, undisturbed watersheds. However, since August 1966, the nitrate concentration has almost continuously exceeded, and at times almost doubled, the maximum concentration recommended for drinking water (Public Health Service, 1962). Thus the deforestation treatment resulted in significant pollution of the drainage stream from the ecosystem.

The high nutrient concentrations, plus the increased amount of solar radiation (due to the absence of forest canopy) and higher temperature in the stream, resulted in significant eutrophication. A dense bloom of *Ulothix zonata* has been observed in the stream during the summers since deforestation. In contrast, the undisturbed watershed streams are essentially devoid of algae of any kind. This represents a good example of how an overt change in one component of an ecosystem may alter the structure and function, often in unexpected manners, in another part of the same ecosystem or in another interrelated

HUBBARD BROOK ECOSYSTEM STUDY
BASED ON TWO YEARS OF DATA

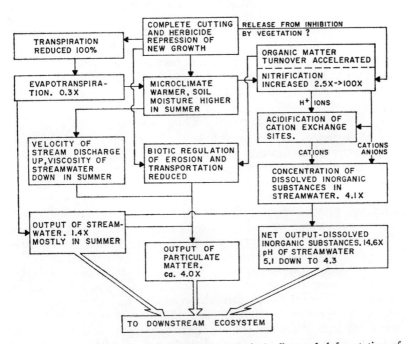

FIGURE 14. A summary of some of the ecological effects of deforestation of Watershed 2 in the Hubbard Brook Experimental Forest. Based upon data obtained during 1966-1968.

ecosystem. Unless these ecological interrelationships are understood, naive management practices can produce unexpected and possibly widespread deleterious results. Several of the various ecological effects and interactions resulting from the deforestation are summarized in Fig. 14.

For example, the tremendous geologic output from a deforestation operation such as this (even though our experiment was excessively severe because vegetation regrowth was inhibited) may be the geologic input for some lake ecosystem. Obviously this could have very serious ecological implications for the lake ecosystem. We have undertaken the study of a lake ecosystem within the Hubbard Brook drainage to evaluate the important biogeochemical and energy linkages between the various ecosystems.

The application of herbicides to the deforested watershed now has been stopped, and we anticipate that vegetation will become re-established. Developmental studies of the revegetation in connection with the study of nutrient cycling using the watershed ecosystem concept should provide some very valuable ecological information. Ecosystem recovery, including the maintenance or re-establishment of homeostatic capacities, is of vital interest to managers and conservationists. We need to know how, and at what speed, ecosystems can recover or regain ecological balance after serious external manipulations.

Summary Thoughts

Wise use of our resource base demands accurate knowledge of the structure and function of ecosystems. Coupled with our model, the small watershed ecosystem concept is useful for simplifying and quantifying the complex interconnections of nutrient cycles within and between natural ecosystems. Quantitative measurements of such nutrient cycles provide major guidelines for the development of sound ecosystem management procedures. Ecologically it is important to protect the vital functions of natural ecosystems. As Leopold (1949) clearly suggested, any sound conservation ethic should be founded upon this premise. Man's manipulations may cause serious imbalances in the ecological function of natural ecosystems. However, with a clear understanding of the functional interrelationships of the ecosystem he may be able to substitute for some of the natural ecosystem's ability to conserve nutrients, while still extracting products desirable to himself. As a simple example, since bark is relatively rich in nutrients, lumbering operations that strip the bark from logs within the ecosystem rather than at some distant processing plant may act to conserve nutrients within the ecosystem. Preliminary estimates at Hubbard Brook indicate that 36 percent of the total calcium in the forest vegetation is incorporated in the stem bark of trees.

Nutrient cycling is closely geared to all components of the ecosystem; decomposition is adjusted to nutrient uptake, uptake is adjusted to decomposition, and both influence chemical weathering. Conservation of nutrients within the ecosystem depends upon a functional balance within the intrasystem cycle of the ecosystem. The uptake of water and nutrients by vegetation is critical to this balance. Removing the transpiration stream from a forest ecosystem seriously affects the homeostatic capacity of that ecosystem to conserve nutrients.

The deforestation study showed that the retention of nutrients within the ecosystem is dependent on constant and efficient cycling between the various components of the intrasystem cycle, i.e., organic, available nutrients, and soil and rock mineral compartments (Fig. 11). Blocking of the pathway of nutrient uptake by destruction of one subcomponent of the organic compartment (i.e., vegetation) leads to greatly accelerated export of the nutrient capital of the ecosystem. From this we may conclude that one aspect of homeostasis of the ecosystem, the maintenance of nutrient capital, is dependent upon the undisturbed functioning of the intrasystem nutrient cycle, and that in this ecosystem no mechanism acts to greatly delay loss of nutrients following sustained destruction of the vegetation.

ACKNOWLEDGMENT: This paper is contribution No. 24 of the Hubbard Brook Ecosystem Study. Financial support has been provided by National Science Foundation Grants Nos. GB-1144, GB-4169, GB-6742, GB-6757, GB-14325, and GB-14289. Published as a contribution to the U. S. Program of the International Biological Program and the International Hydrological Decade. The field study was done through the cooperation of the Northeastern Forest Experiment Station, Forest Service, U. S. Department of Agriculture, Upper Darby, Pennsylvania. We especially acknowledge the individual contributions of John S. Eaton, Donald W. Fisher, Noye M. Johnson, and Robert S. Pierce to the Hubbard Brook Ecosystem Study. We thank G. Evelyn Hutchinson and N. M. Johnson for their comments on the manuscript.

Literature Cited

Alexander, M. 1967. *Introduction to Soil Microbiology*. Wiley and Sons, Inc., New York.

Birge, E. A., and C. Juday. 1934. Particulate and dissolved organic matter in inland lakes. Ecol. Monogr., *4*:440-474.

Bormann, F. H., and G. E. Likens. 1967. Nutrient cycling. Science, *155*:424-429.

Bormann, F. H., G. E. Likens, and J. S. Eaton. 1969. Biotic regulation of particulate and solution losses from a forested ecosystem. BioScience, *19*:600-610.

Bormann, F. H., G. E. Likens, D. W. Fisher, and R. S. Pierce. 1968. Nutrient loss accelerated by clear-cutting of a forest ecosystem. Science, *159*:882-884.

Bormann, F .H., T .G. Siccama, G. E. Likens, and R. H. Whittaker. 1970. The Hubbard Brook Ecosystem study: Composition and dynamics of the tree stratum. Ecol. Monogr., *40*:373-388.

Broughton, W. A. 1941. The geology, ground water, and lake basin seal of the region south of the Muskellunge Moraine, Vilas County, Wisconsin. Trans. Wis. Acad. Sci. Arts Lett., *33*:5-20.

Cantlon, J. E. 1969. Confrontation or cooperation in the cornfield. Ecology, *50*:535.

Cole, L. C. 1958. The ecosphere. Scientific American, April, pp. 83-92.

Cooper, C. F. 1969. Nutrient output from managed forests. In *Eutrophication: Causes, Consequences, Correctives,* pp. 446-463. National Academy of Sciences, Washington, D. C.

Crisp, D. T. 1966. Input and output of minerals for an area of Pennine moorland: the importance of precipitation, drainage, peat erosion, and animals. J. Appl. Ecol., *3*:327-348.

Einsele, W. 1936. Ueber die Beziehungen des Eisenkreislaufes zum Phosphor-kreislauf im eutrophen See. Arch. Hydrobiol., 29:664-686.

Fisher, D. W., A. W. Gambell, G. E. Likens, and F. H. Bormann. 1968. Atmospheric contributions to water quality of streams in the Hubbard Brook Experimental Forest, New Hampshire. Water Resources Res., 4:1115-1126.

Forbes, S. A. 1887. The lake as a microcosm. Reprinted 1925. Bull. Illinois Nat. Hist. Surv., 15:537-550.

Gambell, A. W., and D. W. Fisher. 1966. Chemical composition of rainfall, eastern North Carolina and southeastern Virginia. U. S. Geol. Survey Water-supply Paper, 1535K:1-41.

Hart, G., R. E. Leonard, and R. S. Pierce. 1962. Leaf fall, humus depth, and soil frost in a northern hardwood forest. Northeastern Forest Experiment Sta., Upper Darby, Pa., Res. Note No. 131, 3 pp.

Holstener-Jørgensen, H. 1960. Indfygning of jord i en plantages vestrand. Forstl. Forsøksu. Danm., 26:391-397.

Hornbeck, J. W., and R. S. Pierce. 1969. Changes in snowmelt runoff after forest clearing on a New England watershed. Proc. Ann. Eastern Snow Conf., pp. 104-112.

Hornbeck, J. W., R. S. Pierce, and C. A. Federer. 1970. Streamflow changes after forest clearing in New England. Water Resources Res., 6:1124-1132.

Hutchinson, G. E. 1952. The biogeochemistry of phosphorus. In L. G. Wolterink [ed.], The Biology of Phosphorus, pp. 1-35. Michigan State College Press, East Lansing.

Hutchinson, G. E. 1957. A Treatise on Limnology. Vol. I. Wiley and Sons, Inc. New York.

Johnson, N. M., G. E. Likens, F. H. Bormann, and R. S. Pierce. 1968. Rate of chemical weathering of silicate minerals in New Hampshire. Geochim. Cosmochim. Acta, 32:531-545.

Johnson, N. M., G. E. Likens, F. H. Bormann, D. W. Fisher, and R. S. Pierce. 1969. A working model for the variation in streamwater chemistry at the Hubbard Brook Experimental Forest, New Hampshire. Water Resources Res., 5:1353-1363.

Juang, F. H. F., and N. M. Johnson. 1967. Cycling of chlorine through a forested watershed in New England. J. Geophys. Res., 72:5641-5647.

Junge, C. E. 1958. The distribution of ammonia and nitrate in rain water over the United States. Trans. Amer. Geophys. Union, 39:241-248.

Junge, C. E., and R. T. Werby. 1958. The concentrations of chloride, sodium, potassium, calcium and sulfate in rainwater over the United States. J. Meteorol., 15:417-425.

Kormondy, E. J. 1969. Concepts of Ecology. Prentice-Hall, Inc., Englewood Cliffs, New Jersey.

Leopold, A. 1949. A Sand County Almanac. Oxford Univ. Press, New York.

Likens, G. E., F. H. Bormann, N. M. Johnson, and R. S. Pierce. 1967. The calcium, magnesium, potassium and sodium budgets for a small forested ecosystem. Ecology, 48:772-785.

Likens, G. E., and F. H. Bormann. 1971. Chemical analyses of plant tissues from the Hubbard Brook Ecosystem in New Hampshire. Yale Univ., School of Forestry Bull. No. 79. 25 pp.

Likens, G. E., F. H. Bormann, and N. M. Johnson. 1969. Nitrification: importance to nutrient losses from a cut-over ecosystem. Science, 163:1205-1206.

Likens, G. E., F. H. Bormann, N. M. Johnson, D. W. Fisher, and R. S. Pierce. 1970. The effects of forest cutting and herbicide treatment on nutrient budgets in the Hubbard Brook watershed-ecosystem. Ecol. Monogr., 40:23-47.

Likens, G. E., F. H. Bormann, R. S. Pierce, and D. W. Fisher. 1971. Nutrient-hydrologic cycle interaction in small forested watershed-ecosystems. Proc. Colloque sur la Productivite des ecosystemes Forestiers dans le Monde, Bruxelles. (In press.)

Lunt, H. A. 1932. Profile characteristics of New England forest soils. Connecticut Agric. Expt. Sta., Bull. No. 342:743-836.

Neil, J. H., M. G. Johnson, and G. E. Owen. 1967. Yields and sources of nitrogen from several Lake Ontario watersheds. Proc. Tenth Conf. Great Lakes Res., pp. 375-381.

Nye, R. H., and D. J. Greenland. 1960. The soil under shifting cultivation. Commonwealth Bureau of Soils, Harpenden, England, Tech. Bull. No. 51, 156 pp.

Odum, E. P. 1959. Fundamentals of Ecology (2nd ed.). W. B. Saunders Co., Philadelphia.

Odum, E. P. 1962. Relationships between structure and function in the ecosystem. Japanese J. Ecol., 12:108-118.

Pierce, R. S. 1967. Evidence of overland flow on forest watersheds. Proc. Internat. Symposium Forest Hydrology, pp. 247-252. Pergamon Press, Inc., New York.

Pierce, R. S., J. W. Hornbeck, G. E. Likens, and F. H. Bormann. 1970. Effects of elimination of vegetation on streamwater quantity and quality. Proc. IASH Symposium on the Results of Research on Representative and Experimental Basins, New Zealand, pp. 311-328.

Public Health Service. 1962. Drinking water standard. Public Health Service Publ. 956. Washington, D.C.

Ruttner, F. 1963. Fundamentals of Limnology (3rd ed.). Translated by D. G. Frey and F. E. J. Fry. Univ. of Toronto Press, Toronto.

Sawyer, C. N. 1947. Fertilization of lakes by agricultural and urban drainage. J. New England Water Works Assn., 61:109-127.

Siccama, T. G., F. H. Bormann, and G. E. Likens. 1970. The Hubbard Brook Ecosystem study: productivity, nutrients, and phytosociology of the herbaceous layer. Ecol. Monogr., 40:389-402.

Sylvester, R. O. 1961. Nutrient content of drainage water from forested, urban and agricultural areas. Algae and Metropolitan Wastes, U. S. Public Health Service, SEC TR, W61-3:80-87.

Tamm, C. O., and T. Troedsson. 1955. An example of the amounts of plant nutrients supplied to the ground in road dust. Oikos, 6:61-70.

U. S. Forest Service. Northeastern Forest Experiment Station. 1964. Hubbard Brook Experimental Forest. Northeastern For. Exp. Sta., Upper Darby, Penn., 13 pp.

Westlake, D. F. 1963. Comparisons of plant productivity. Biol. Rev., 38:385-425.

Whitehead, H. C., and J. H. Feth. 1964. Chemical composition of rain, dry fallout, and bulk precipitation at Menlo Park, California, 1957-1959. J. Geophys. Res., 69:3319-3333.

Discussion

QUESTION: What happened to the herbicides that were used in the deforested watershed treatment?

DR. LIKENS: We monitored herbicides in stream water throughout the course of the experiment. The major herbicide application employed Bromocil, which is supposed to be a slowly decomposing, long-acting herbicide. We used this herbicide because it is effective against higher plants, it contains the smallest amount of nutrients involved in the study, and because we wanted to eliminate vegetation regrowth for a long period of time. The Bromocil was broadcast aerially at a rate of 28 kg/ha. It appeared for a year and a half in the drainage water in concentrations of about 0.25 ppm, although concentrations as high as 1.8 ppm were observed. In terms of the herbicide function this means it probably was functioning quite well and decomposing quite slowly. Smaller amounts of 2, 4, 5-T were applied individually with backpack sprayers, and only on stump sprouts, etc. The concentration of 2, 4, 5-T in stream water was less than one ppb for the entire period following application.

Let me make one more point here which perhaps I didn't stress sufficiently earlier: This experiment represents a rather severe manipulation. Normally good forestry practice would not try to inhibit vegetation regrowth for a long period as we did, and this makes the direct application of our results somewhat compromised. However, we have looked at a number of other watershed areas, one just across the ridge-crest from Hubbard Brook and a few others within the White Mountains, that have been cut by commercial loggers. We found the same pattern as at Hubbard Brook, high nitrate concentration and *Ulothrix zonata* in the streams. In one of these watersheds, which had been cut several years prior to our analyses, the stream-water nitrate concentrations greatly exceeded the normal stream-water concentrations for the region.

QUESTION: In regard to, say, nitrates, what would have happened to your runoff if the forest had not had a water impervious base?

DR. LIKENS: It is fair to assume that we would have observed the same very large increases in the nitrate concentrations in the drainage water. The problem would have been that we wouldn't have been able to put together any quantitative budgets for the system, i.e., some amount would have been lost as deep seepage and would not have been measurable.

QUESTION: What do you think about the practice of leaving "filter-strips" of vegetation along a stream in clear-cutting operations to prevent degradation of the stream? Is this an effective method in your estimation?

DR. LIKENS: We now have studies under way to answer that question for our area in a quantitative way. We have two manipulations planned. One is to put into effect a strip-cutting operation in which horizontal strips will be cut across the watershed all the way from the mouth of the watershed to the top. In addition, a narrow filter strip, 25 m, will be left along either side of the stream channel. This was designed with the advice of forest management personnel and forest hydrologists. We hope to determine whether it is possible to "give the hydrologist some excess water output, allow the forest manager to realize a timber output, yet conserve the nutrients within the ecosystem." We think there is a good possibility that the nutrients may be held up in these strips rather than being flushed out of the system. The mature natural ecosystem is a very conservative system and has a strong homeostatic capacity to maintain a balance of vital nutrients like nitrates. When you eliminate this homeostatic capacity, there seems to be nothing else in the system that can take over the functional two-way exchange between available nutrients and organic matter.

The other operation planned is a strictly commercial logging operation with roads, skid trails, and the whole bit.

Energy Flux in Ecosystems

FRANK B. GOLLEY

Institute of Ecology and Department of Zoology
University of Georgia, Athens, Georgia 30601

STUDIES OF THE ENERGY DYNAMICS OF ecosystems and populations have increased greatly over the past decade, resulting in a reasonably firm theoretical base for the subject and a growing body of data. The evolution of these studies, like most subjects of inquiry, has resulted in an intensification of focus and a deeper analysis of the details of energy flux of ecological systems. This discussion will take an alternate point of view. My object will be to examine the energy flux of the whole biosphere, the use of energy for work in the most significant ecosystems, and finally, the implications of energy flux studies to broader questions of the environmental crisis now facing mankind. Such a broad-brush approach might be appropriate in a symposium on ecosystem structure and function in 1970.

The Energy Flux to the Planet

It is well known that the energy which drives the ecosystems on the planet Earth comes from the sun. At the outer limits of our atmosphere, 1.94 gram calories of solar energy per square centimeter are received per minute. Not all of this energy reaches the surface of the earth; 35 percent is reflected and 17.5 percent is absorbed by the atmosphere and clouds. This leaves only about 47.5 percent of the solar radiation effective at the level of the biosphere (Linton, 1965). Naturally, there is wide local variation in solar radiation input to the earth depending upon the cloud cover, turbidity of the atmosphere, and other factors.

How is this energy used? About 30 percent is reflected as long wave radiation, 49 percent is used in the evaporation and condensation of water, and 21 percent is exchanged by conduction of heat to the air through movement of wind over the surface of land and sea (Gates, 1962). The energy expended in evaporation and condensation and in heat conduction drives the circulation of the atmosphere and oceans, which truly tie the biosphere together into a one-world ecological system. Differential absorbtion of solar energy at the surface results in the flow of energy through atmospheric movements from regions of high energy absorbtion to areas of low energy absorbtion. Atmospheric circulation acting on the oceans produces the water movements that erode coastlines. Evaporation from the oceans drives the hydrologic cycle, in which water is evaporated from the oceans, precipitated onto the land and returns to the oceans in rivers, effecting the downcutting of the continents in the process. These are the patterns of wind and water which make up the environments in which living organisms act out their existence.

Influx to the Biosphere

The living portion of the planetary ecological system requires energy to maintain the thermodynamically unstable condition of life. It is common knowledge that this energy is supplied by solar energy captured by green plants through the process of photosynthesis. In this process light energy is absorbed by chlorophyll and other pigments in the chloroplasts. The photosynthetic mechanism consists of a number of reaction systems in which high energy phosphate and hydrogen donors are produced by photochemical reactions and are used in carbon reduction. The product of the photosynthetic reduction of carbon dioxide may be a carbohydrate but also may be other organic substances, depending upon the physiological state of the cell. The accumulation of these products is termed gross primary production. The organic substances containing potential chemical energy may be used in the catabolic processes of the plant, in which case there is a loss of energy from the plant. This loss is conventionally termed respiration in the ecological literature. The organic substances may also be stored or excreted. The energy content of these substances can be determined by burning them in a bomb calorimeter. The rate of storage is termed net productivity, and the total amount of storage over the growing season, including that dying and eaten during this period, is the net primary production. The total energy flow (gross primary production) into the vegetation includes the net primary production plus the energy of respiration. Since net production is what is commonly measured in the field and gross production is calculated from net values, we will begin our analysis with net production.

Net Primary Production

The net primary production of terrestrial vegetation can be determined reasonably accurately by harvesting the plants at the end of their growing season and determining their dry weight and energy content, with suitable corrections for unharvested roots and material lost by leaf fall and by grazing. Net primary production of aquatic plants is more difficult to estimate. The plankton consists of small organisms containing a few cells, which have a relatively short life span. The techniques used to measure photosynthesis in these small organisms are difficult to compare. Also, plankton produce extracellular products which are seldom included in production estimates. Fogg (1968) suggests that aquatic net primary production values may have to be increased 25 percent to account for production of extracellular organic materials. For these reasons, there are numerous estimates of terrestrial net primary production but few estimates of planktonic net primary production. Recently such estimates have been collected by Lieth (1962), Westlake (1963), Ovington (1965), Tadaki *et al.* (1965), and Kira and Shidei (1967).

The biosphere net primary production can be estimated from information on the average net primary production and on the areal extent of the major world ecosystems. Following the conventions of classical ecology, the biosphere has been divided into the ecosystems listed in Table 1. Next, the area occupied by each ecosystem was determined. Actually there is considerable disagreement about the areal extent of these systems (Table 2) due to differences in accuracy of maps and to different definitions of the ecosystems themselves. A median area was chosen for each ecosystem for the purposes of this paper (Table 2). The net primary production estimates were adopted from Whittaker (1970), derived from the reviews listed above. The data on net primary production for an ecosystem vary widely because of differing environmental conditions over space and time. For most systems the distribution of data are skewed to the left, indicating that most stands have a relatively lower productivity, while a few have a very high productivity. The estimates in Table 1 are rounded median estimates for all conditions within a specified ecosystem type. Multiplication of the areal extent of an ecosystem by the median net primary production per square meter then gives the biosphere net production.

Comparison of the estimates of world net primary production shows that production by the oceans is little more than one-half the terrestrial production (Table 1), and that the tropical forests and open oceans are the most productive ecosystems, but for different reasons. Tropical forests have a high rate of production but a relatively small area; the open oceans, on the other hand, have a small rate of production but a large area. There are several reasons why aquatic systems have a low production rate. In water, available light may be reduced by the water itself and by suspended or dissolved material. Further,

Table 1. Estimates of primary production in various ecosystem types

Ecosystem	Areal extent*	Average caloric value†	Biosphere				Ecosystem			
			Gross production‡		Net production		Gross production‡		Net production§	
	$10^6 km^2$	kcal/g	$10^{12} kg/yr$	$10^{18} kcal/yr$	$10^{12} kg/yr$	$10^{18} kcal/yr$	$g/m^2/yr$	$kcal/m^2/yr$	$g/m^2/yr$	$kcal/m^2/yr$
Tropical forest	20	4.2	133	55.9	40.0	16.8	6,700	28,140	2,000	8,400
Temperate forest	6	4.6	26	12.0	7.8	3.6	4,300	19,780	1,300	5,980
Boreal forest	10	4.7	27	12.7	8.0	3.8	2,700	12,690	800	3,760
Tropical Savanna	15	4.0	18	7.2	10.5	4.2	1,200	4,800	700	2,800
Grassland	25	4.0	13	5.2	7.5	3.0	500	2,000	300	1,200
Desert	25	4.0	3	1.2	1.8	0.7	120	480	70	280
Cultivated land	15	4.3	16	6.9	9.8	4.2	1,100	4,730	650	2,795
Tundra	10	4.7	3	1.4	2.0	0.9	300	1,410	200	940
Snow and ice	15	0.0	0	0.0	0.0	0.0	0	0	0	0
Other+	7	4.0	2	8.0	1.4	0.6	300	1,200	200	800
TOTAL LAND	148		241	110.5	88.8	37.8				
Open ocean	332	5.0	69	34.5	41.5	20.7	200	1,000	125	625
Continental shelf	27	5.0	16	8.0	9.5	4.7	600	3,000	350	1,750
Estuaries	2	5.0	7	3.5	4.0	2.0	3,300	16,500	2,000	10,000
TOTAL OCEAN	361		92	46.0	55.0	27.4				
TOTAL BIOSPHERE	509		333	156.5	143.8	65.2				

* From Table 2.
† Based on data from Golley (1961), Ovington and Heitkamp (1960), Long (1934), and Cummins (1967).
‡ Calculated from net production values by assuming net as 30 percent of gross in tropical, temperate, and boreal forests, and as 60 percent of gross in other vegetation.
§ From Whittaker (1970).
+ Includes lakes, rivers, scrub land, and extreme desert.

Table 2. COMPARISON OF ESTIMATES OF THE AREAL EXTENT OF WORLD ECOSYSTEMS (AREA IN 10^6 KM^2)

Ecosystem	Authority					Area used in this report
	Whittaker (1970)	Bowen (1966)	Schmidt (1966)	Lieth (1964)	Stamp* (1960)	
Forests	....	44	45	43.3	54.4	
Boreal	12	10	4	10.0	8.6	10
Temperate	18	4.9	6	4.9	26.8	6
Tropical	20	14.7	20	14.7	19.0	20
Other	....	14.4	15	14.0	......	---
Savanna	15	......	....	......	12.9	15
Grassland	....	27	37	25.7	22.5	25
Desert	18	33	22	52.0	25.2	25
Cultivated	14	23	10	14.0	7.9	15
Tundra	....	11	9	......	9.9	10
Snow and ice	....	11	19	12.7	12.4	15
Other	....	....	....	....	......	7
TOTAL	149	148	142	148	145	148

* Based on world soil groups.

nutrient salts necessary for plant growth may be limited in water because only those adjacent to the cells are accessible for uptake. In contrast, roots of terrestrial plants can extend into a large area of substrate for necessary nutrients.

Gross Primary Production

Next it is necessary to convert the estimates of net primary production to gross primary production by adding the material expended in plant maintenance. There are numerous estimates of plant respiration under field conditions. While there are methodological problems in comparing respiration data and extrapolating data gathered for single plant parts to the whole vegetation, a sample of the available data nevertheless may provide a basis for correcting net primary production. Most crops and natural vegetation, including phytoplankton, appear to use about 40 percent of the gross primary production in plant respiration (Fig. 1). However, mature temperate and tropical forests seem to differ from this general trend—these ecosystems use about 70 percent of the gross photosynthesis in maintenance. Respiration of forest trees increases with tree size (Kira and Shidei, 1967), and it is probable that the difference between mature forests and other kinds of vegetation shown in Figure 1 is due to a difference in the quantity of woody tissue in the system.

If we accept the estimates of plant respiration as a percent of gross photosynthesis shown in Figure 1, then we can calculate gross primary production from the estimates of net primary production in Table 1. I have assumed that in tropical, temperate, and boreal forests, 30 percent of the gross primary production is net primary production and in other ecosystems 60 percent of the gross primary production is net production. Gross primary production must be multiplied by the mean energy content of the vegetation to obtain a value for the energy content of the biosphere. While plant material is considered to have a mean caloric value of between 4.0 and 4.5 kcal/g dry weight, the energy values of the vegetation, based on random samples of the entire vegetation complex, do differ significantly. Vegetation in the tropics has a lower overall energy content than vegetation in polar regions (Golley, 1969). Table 1 reflects these differences. Comparison of gross primary production of world ecosystems shows that two ecosystems are most significant in energy capture. These are the tropical forest (responsible for 36 percent of the total energy flux into the biosphere) and the open oceans (responsible for 22 percent). Somewhat less than three times as much solar energy is converted to chemical energy by land plants as by ocean plants.

With these estimates of gross primary productivity, we can compare the flux of energy into the biosphere to the amount of solar energy available at the earth's surface. According to Linton (1965), 3.7×10^{18} kcal are received by the earth's disc each day. Since only 47.5 percent reaches the earth's surface, the available solar energy equals 1.8×10^{18} kcal/day or 657×10^{18} kcal/yr. Gross primary production of the biosphere is 1.6×10^{18} kcal/yr, or 0.24 percent of the available solar energy. While the numbers and assumptions used in these calculations will be revised further as detailed analyses of ecosystems flux are completed, especially through the International Biological Program, the general conclusion is that only a small fraction of the available solar energy is captured by the biosphere.

Limits to Primary Production

We might digress for a moment to consider why so little of the available solar energy is converted to chemical energy by the vegetation. While this question is exceedingly complicated and merits an in-depth analysis, we can identify at least four phenomena which influence the rate of energy conversion by plants.

First, the process of photosynthesis operates theoretically at an efficiency of about 20 percent (Bonner, 1962). Bonner estimates that about 10 quanta are required to reduce one CO_2 molecule. Since 10 quanta supply about 520 kcal of energy per mole of CO_2, and one mole of CO_2 stores about 105 kcal, the efficiency of the process is 0.20 (105/520 = 0.20).

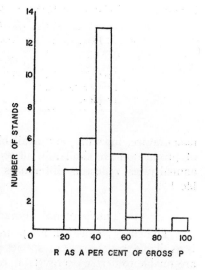

FIGURE 1. The distribution of stands of vegetation as a function of the percentage respiration energy is of gross primary production. Two peaks are identified. One at 40-50 percent and the second at 70-80 percent. This latter peak represents mature tropical and temperate forests.

Second, photosynthesis does not increase directly with light intensity. In many land plants photosynthesis increases with light intensity to about 1/10 to 1/20 full sunlight, then the rate becomes independent of further increase of light until high light intensity is reached. At this point photosynthesis may be adversely affected.

Third, there is a limit to the amount of chlorophyll or leaves that can be distributed over a surface. In terrestrial systems the area of leaves over the area of ground surface is called the leaf-area index (LAI). Many crop plants reach a maximum production at an LAI of 4 or 5 m² leaves/ m² of ground. If the LAI goes above this level competition for light is such that the overall productivity of the vegetation is reduced. The LAI of natural vegetation may be much greater than in crops; tropical forests may have LAI as high as 20 m²/m² but this seems to be near the natural limit.

Finally, photosynthesis utilizes only those wave lengths of light absorbed by chlorophyll; for example, chlorophyll has sharp absorbtion peaks at 4,300 Å and at 6,700 Å (Gates, 1962). Generally, chlorophyll is sensitive to only about one-half of the solar radiation received at the earth surface. These four factors, the quantum efficiency of photosynthesis, the light saturation of chlorophyll, the distribution of chlorophyll over the earth or water surface, and the utilization of only a part of the light spectrum, all contribute to the low observed efficiency of influx of solar energy to the biosphere.

Given these limits to the production process, annual production of vegetation also is strongly influenced by environmental factors such as

Table 3. MAXIMUM DAILY RATES OF PRIMARY
PRODUCTION IN DIFFERENT ECOSYSTEMS

Ecosystem	Xeric	Mesic	Hydric
	g/m^2	g/m^2	g/m^2
Arctic	1.2	11.1	3.7
Temperate	1.2	9.5	28.0
Tropical	....	17.0	16.0

temperature, moisture, and available nutrients. Maximum daily rates of production are similar in all types of vegetation (Table 3), while annual rates, reflecting differences in environment, vary greatly (Table 1).

The Internal Dynamics of Ecosystems

The energy stored in net primary production by the green plants is available to the array of species populations in the ecosystem which are unable to derive energy from other sources. These populations include the animals, fungi, and certain bacteria. In many ecosystems this ensemblage contains 2,000 or more species, and for this reason a detailed reconstruction of the ecosystem from a study of each of its constituent species populations is impractical. In very general terms, the dependent species populations can be divided into those ultimately dependent upon living plant material and those dependent upon dead plant material (Fig. 2). Naturally, there is some crossover between these broad categories—a bird may eat a beetle consuming dead wood and an orthopteran feeding on green leaves—but the categories distinguish between two broadly different paths of energy flow which have a different significance to the system. Living plant feeders influence the production machinery of the system directly by feeding on green leaves, stems, or living roots; dead plant feeders, on the other hand, influence the living plants indirectly by releasing nutrients tied up in the dead plant parts for uptake through the active roots. What is the pattern of distribution of energy to these two pathways?

Flow to Herbivores

Estimates of the quantity of net primary production naturally consumed by living plant feeders (herbivores) are given in Table 4. Obviously, the values have a wide distribution since it is well known that herbivores can and do defoliate the plant community under certain conditions. However, the data are strongly skewed to the left, especially in forests and grass and herbaceous systems where the major herbivores are invertebrates. The degree of skewness is less strong for

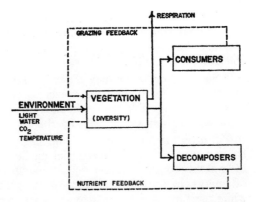

FIGURE 2. A simplified model of the interactions of vegetation, consumers, and decomposers in an ecosystem.

grassland with abundant vertebrate herbivores, as in Africa, and in certain aquatic systems such as springs. From these data we conclude that in most forests, and many nonforest ecosystems, less than 10 percent of the energy in net primary production flows to herbivores, while in grasslands and some aquatic systems less than 50 percent of the energy flows on this pathway. These observations may have several explanations.

Table 4. NET PRIMARY PRODUCTION OR ORGANIC INPUT TO THE SYSTEM
CONSUMED BY HERBIVORES

Ecosystem	Organic input consumed	Authority
	Percent	
Tropical forest	8.5	Misra (1968)
Tropical forest	7	Odum and Ruiz-Reyes (1968)
Temperate forest	1.5–2.5	Bray (1964)
Temperate forest	1.5	Reichle and Crossley (1967)
Temperate forest	3.4–9.2	Kazmarek (1967)
Temperate forest	40	Andrezejewski (1967)
Old-field	12	Odum, Conell, and Davenport (1962)
Lespedeza field	0.4–1.4	Menhinick (1967)
Old-field	1.0	Wiegert and Evans (1967)
Salt marsh	8	Teal (1962)
Meadow	25–30	Andrezejewski (1967)
Tanganyika grassland	28	Wiegert and Evans (1967)
Indian grassland	49	Misra (1968)
Uganda grassland	60	Wiegert and Evans (1967)
Cultivated potatoes	12–20	Trojan (1967)
Silver Spring	38	Odum (1957)
Cone Spring	24	Tilly (1968)
Ocean, Georges Bank	5	Clarke (1946)

First, forests route some net production to stems, branches, and roots, and herbivory of these parts is either small or poorly known. In contrast, the mass of other vegetation may be succulent and fully available for consumption.

Second, vegetation may differ in its relative ability to resist grazing. Where the chloroplasts are enclosed in elaborate leaf systems, these may be less able to tolerate heavy destruction than systems where chloroplasts are in simple cells, floating in a watery medium.

Finally, characteristics of the herbivores (Odum, 1956), their habitat in an aquatic or terrestrial medium (Misra *et al.*, 1968), and their ability to graze in tree tops all may play a role in the distribution of energy in the system. Regardless of the details, it appears that while most vegetation can tolerate some degree of defoliation, apparently the tax regularly imposed by herbivores on the production machinery of the system is relatively low.

Flow to Decomposers

The energy flowing to the decomposer organisms is used in the work of breaking down the complicated biological structure built through the production process. The species populations exhibit different degrees of refractiveness to decomposition, so that the extent and amount of materials released to the substrate and available for uptake vary with the species populations in the ecosystem as well as with the environment. Energy also flows to the decomposer portion of the ecosystem from the herbivore food web as organisms die and are decomposed, or as material is voided from the body as fecal or excretory wastes or as secretions. Further, not all the organic material dying in a system may be decomposed. Dead organic matter may be moved by wind and water, with the result that dependent ecosystems, such as some streams and rivers, obtain most of their energy as import of organic material. In these latter systems primary producers may be unimportant or absent, and the system must be considered together with the system from which it obtains organic material. This is part of the rationale for considering entire watersheds as systems rather than focusing on the land or rivers singly. Dead organic matter may also be stored and accumulated.

The two categories, herbivores and decomposers, conceal complicated patterns of anastamosing populations. The transfer of energy between these populations is poorly known in detail. If we are willing to organize the populations into abstract categories such as herbivores, carnivores, top carnivores, etc., then probably 10 percent or less of the energy of a level is transferred to the next level. More data are required for detailed analysis of the internal dynamics of the consumers.

The Work of Ecosystems

In elementary physics books energy is described as the capacity to do work. We might ask, what work is accomplished in the ecosystem with the energy converted through photosynthesis? Lotka's (1924: 335) answer to this question was that most of the energy is used in system maintenance:

> The picture we must keep before us, then, is that of a great world engine or energy transformer composed of a multitude of subsidiary units, each separately, and all together as a whole, working in a cycle. It seems, in a way, a singularly futile engine, which, with a seriousness strangely out of keeping with the absurdity of the performance, carefully and thoroughly churns up all the energy gathered from the source. It spends all its work feeding itself and keeping itself in repair, so that no balance is left over for any imaginable residual purpose. Still, it accomplishes one very remarkable thing; it *improves* itself as it goes along, if we may employ this term to describe those progressive changes in its composition and construction which constitute the evolution of the system.

While we may want to quibble with Lotka's statement, his point that the major work of the ecosystem is maintenance and self-repair is essentially correct. But as a consequence of maintenance, ecosystems also act upon the geosphere, retarding erosive forces and augmenting chemical breakdown of minerals, and upon the atmosphere through the uptake and release of oxygen, carbon dioxide, and other gases and substances. All of these actions are set into motion by the energy flux through the ecosystem and are appropriately considered in this context.

Maintenance and Adaptation of Ecosystems

Ecosystems tend to evolve toward a state where, as Lotka says, all the energy is used in system maintenance. We theorize that once this state of equilibrium is reached, the system changes only very slowly through the continued evolution of its member species. Equilibrium is defined in both structural and dynamic terms. The structure of the equilibrium ecosystem remains constant over time, while all of the gross primary production is expended in maintenance of the system. Some maintenance is for the vegetation and some is for the consumer and decomposer populations. There is no storage of energy in the equilibrium system. Thus, the energy cost of maintenance of an equilibrium system is equal to its gross primary production. Estimates of maintenance expense based on net production data corrected for energy costs (Fig. 1) clearly show that forests have large maintenance costs, while nonforest systems have a lower expense (Table 1).

Another aspect of maintenance concerns the influence of ecosystems on the geosphere and on the atmosphere. It is well known that the vegetation retards erosion by reducing the effect of rain on the soil, by holding the soil through root action, by the formation of drainage channels in the soil, and by other factors. While the biomass of vegetation per unit area increases with an increase in precipitation, the yield of sediment from a drainage basin also increases with precipitation. Considering these two opposing actions, Langbein and Schumm (1958) have calculated the effect of various ecosystems on relative erosion (Fig. 3). Forests are most effective in reducing erosion, although grasslands may equal forests at certain levels of precipitation. Row crops or small-grain agriculture may increase relative erosion 10 to 100 times over that of forests. If this example is generally applicable to other regions, we can visualize the influence of the vegetation on the rate of erosion of the continents as part of the work of ecosystems. In the United States the rate of denudation of the land is lowest in the forest region (4.1 cm per 1,000 years) compared to the Colorado drainage area consisting of grassland and desert, which has a rate of 16.5 cm per 1,000 years (Judson and Ritter, 1964).

Conversion of forested regions to agriculture can be expected to increase erosion, especially in tropical forest areas. The effect of settlement and agriculture is illustrated by a comparison of the sediment yield of tropical rivers of South America, an area of low settlement, and Asia, an area of high density settlement. The average annual sediment yield of rivers draining rain forest areas in South America is 215 tons/square mile (75 metric tons/km^2), while in Asia it is 1,658 tons/square mile (581 metric tons/km^2; Holeman, 1968). The difference between these estimates of sediment yield is almost a factor of ten, which is similar to the factor Langbein and Schumm (1958) derived comparing relative erosion under crops and forest. The Amazon River discharges 11 to 18 percent of the world's total annual volume of water to the oceans, and it discharges 20 percent of the total sediment input to the oceans. Large-scale reservoir construction or agricultural development on the Amazon would be expected to change these yields.

One of the consequences of the energy flux of ecosystems is the consumption of carbon dioxide and production of oxygen through the photosynthetic process. Over geological time, carbon dioxide has been deposited as carbonate or organic carbon in sedimentary rocks. Mason (1966) estimates that the carbon dioxide in sedimentary rock is roughly 100 times that in the present atmosphere, hydrosphere, and biosphere. Hutchinson (1954) points out that the differential solubility of oxygen and carbon dioxide in sea water has resulted in a greater

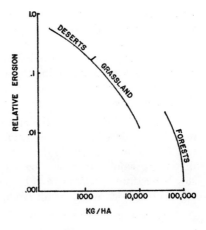

FIGURE 3. The relationship between mass of vegetation and relative erosion, *from* Langbein and Schumm (1958).

quantity of carbon dioxide in the ocean. These authors state that probably most of the oxygen in our atmosphere is the result of photosynthesis.

Oxygen is also used in oxidation of rock at the earth's surface and in oxidation of volcanic gases. Apparently during periods of orogeny, oxygen levels in the atmosphere fell due to oxidation of newly exposed rock (Rutten, 1966). It has been suggested that the periods of evolution of the biota, which are correlated with periods of orogeny, were stimulated by these fluctuations in atmospheric oxygen and carbon dioxide. In any event, we are very much concerned about these interactions today because we may influence the cycles of oxygen and carbon dioxide through release of chemicals into the hydrosphere and atmosphere.

For example, it has been suggested that movement and buildup of foreign chemicals in the oceans might destroy the ocean plankton and cause a change in the oxygen level of the atmosphere. One assumption on which this hypothesis is based is that the sea produces the major part of the oxygen in the cycle. Our consideration of the relative significance of various ecosystems in energy capture (Table 1) suggests that the ocean produces only about one-third as much as the land. If the production capacity of the ocean could be destroyed instantaneously, then about 92×10^{12} kg of production would be eliminated, and assuming that the production of oxygen is about equal to the production of organic matter, then 92×10^{12} kg of oxygen would no longer be released to the atmosphere. Since the free oxygen in the atmosphere is 12×10^{17} kg (Mason, 1966), the decreased input would be about 0.007 percent of the quantity in the atmosphere. In addition, the dead plankton would take up oxygen in their decay. I have been unable to obtain a firm estimate of the standing crop of biomass in the ocean, but if we accept the lower estimates from such fertile places as Georges Bank, Newfoundland, and the English Channel, then there might be about 5 g/m² of ocean. Multiplying this estimate by the area of ocean gives a standing crop of plankton of 2×10^{12} kg. Less than 0.0001 percent of the atmospheric oxygen would be required to oxidize this biomass.

Naturally, this is not the entire story. Since the oxygen in the atmosphere is derived from photosynthesis, destruction of photosynthesis will lead to an inexorable decline in atmospheric oxygen. The point is that there would probably be a certain time lag in response, with a slowly declining percentage of oxygen in the atmosphere as oxygen was taken up in rock weathering. An apocalyptic effect is less probable.

According to recent data of Bolin and Keeling (1963) the level of carbon dioxide in the atmosphere has increased. An increase in carbon dioxide has been reported since about the late 1800's; however, only recently have accurate, comparable data been obtained using modern instrumentation. It has been suggested that increase in the carbon dioxide level in the atmosphere is a cause for concern since this gas produces a greenhouse effect and would increase the heat at the earth's surface, which in turn would cause an increase in melting of the polar ice caps, resulting in an increase in mean sea level. While the complications of such a chain of hypothetical events might make one cautious about taking such a problem very seriously, it may be worthwhile to examine the immediate cause of the increase of atmospheric carbon dioxide since it is related to the energy flux of ecosystems.

The carbon dioxide increase in the atmosphere could come from combustion of fossil fuels, or through decomposition of the organic matter stored in ecosystems which are converted to agricultural systems. Fossil fuel combustion in 1960 was estimated to release 13×10^{15} g carbon dioxide (Bolin and Keeling, 1963). The rate of combustion of fossil fuel has increased as industrial development has expanded (at about 0.7 ppm per pear), and we can anticipate that this increase will continue.

It is quite difficult to evaluate the effect of converting natural ecosystems to agricultural ecosystems. In a temperate forest soil the organic matter content is about 3.3 percent, while the same soil under cultivation may have 1.4 percent organic matter (Giddens, 1957). If we assume a soil depth of 6 inches (15.2 cm) and the weight of an acre furrow slice to be 2×10^6 pounds (9.1×10^5 kg), then the organic matter lost from this soil is about 43×10^6 g/ha or 60×10^6 g of carbon dioxide equivalent.

The increase in acreage of cultivated crops such as cereals, potatoes, yams, cotton, pulses, and oil seeds, according to the 1968 FAO Yearbook, was 138 x 10^6 ha from about 1950 to 1967. If this land lost organic matter at the rate estimated for a Georgia temperate forest soil, then the annual release of carbon dioxide by this process would be about 349 x 10^{12} g or 488 x 10^{12} g of carbon dioxide equivalent. These calculations suggest that conversion of forest to farm lands may have contributed only 1/30 as much carbon dioxide to the atmosphere as burning fossil fuel in the recent past.

While these estimates are very preliminary, they show that the maintenance processes of ecosystems have many ramifications. Maintenance of the atmosphere, hydrosphere, and geosphere cannot be accomplished without maintenance of the biosphere. These are all interacting subsystems which are part of one planetary ecosystem.

Repair

If an equilibrium ecosystem is disturbed and its structure is not changed, we consider the energy expense to counter the disturbance to be maintenance. In contrast, if the disturbance is sufficiently severe so that the system structure is materially altered, then repair processes begin. Repair continues until equilibrium conditions are reestablished. Seldom are repair processes operating over short time periods. If disturbance is extremely severe, repair may require centuries, and, indeed, as far as man is concerned certain highly fragile systems may require periods of geological time for the reorganization of equilibrium ecosystems on the landscape. Nevertheless, in most situations today, man is concerned with repair processes with time spans of no more than a few hundred years. If this analysis is correct, then all of the ecosystems on earth can be arranged into four categories: (1) equilibrium systems; (2) systems undergoing repair; (3) systems where man, through management, is controlling the repair process; and (4) severely disturbed systems where repair processes cannot overcome the influence of adverse environmental factors. Since accidents happen to equilibrium systems on which man depends and man acts as a disturbing agent in the environment, it is important to understand the repair process. What is the energy expense of repair in contrast to the energy required for maintenance? Let us consider this comparison in the tropical forest.

In the tropics, ecological succession, as the repair process is called, proceeds rapidly in the initial stages. Grass, vines, and trees are present the first year after a bare area becomes available for revegetation. In Panama, we found that the leaf surface area characteristic of the mature forest was reestablished by the sixth year after abandonment of subsistence agricultural fields (Duever and Ewel, 1971). The secondary forest is fully established in 8 to 15 years and primary forest only slowly reasserts itself. Estimates of the time necessary to develop equilibrium forest conditions range from 100 to 200 years (Richards, 1964). However, Richards quotes Chevalier (1948) as saying that the forest on the site of Angkor Wat, Cambodia, destroyed five to six centuries ago, still shows certain differences from the virgin forest of the district. If we assume that the full production capacity of the tropical forest system is restored in 20 years and that equilibrium is reached in 200 years, then we can calculate the energy of gross production required for ecosystem repair over this time span. If gross production increases from 1,300 $g/m^2/yr$ to 6,600 $g/m^2/yr$ over the first 20 years and then is maintained at 6,600 $g/m^2/yr$ for the next 180 years, the total production or repair energy expense over the time span will be 1.3 x 10^6 g/m^2 or about 5 x 10^6 $kcal/m^2$. Since the annual maintenance cost of a tropical forest is about 0.028 x 10^6 $kcal/m^2$, the expenditure of energy to repair the system is roughly 200 times the annual maintenance expense.

While some of the energy of gross primary production is used to maintain the successional communities, a portion is stored as structure in the system. Structure may include the wood of trees and shrubs, consumer populations, and improved soil conditions. The amount of gross primary production stored in the system slowly declines to the mature condition where there is no storage. Since the mature rain forest has a stem and leaf biomass of 260 metric tons/ha, the energy lost in destruction of the forest is about 0.118 x 10^6 $kcal/m^2$, assuming the tissue has a caloric value of 4.5 kcal/g dry weight. Thus, it requires about 50 times the energy stored in the mature biomass to reestablish the system.

Odum (1968) has described the small quantities of energy required to disrupt the tropical rain forest in a radiation experiment in Puerto Rico. Energy of disordering work (1.9 $kcal/m^2$) was 1/6200 of the energy of the ordering work (maintenance energy) in this system. The subsistence agriculturalist who cuts down and burns patches of tropical forest for his small farm expends about 3,000 kcal/day for part of about 6 months to cut down the forest. Assuming that it requires about 2 months of one man's labor to clear one hectare of forest, the energy required for disordering the forest is 18 $kcal/m^2$, or about 1/1600 of the annual maintenance cost of the forest. Obviously, the energy required to disrupt a system varies with the means of disorder, the degree of disruption, and the type of system. It would seem important to know the effect of various means of disturbance and identify fragile systems for successful ecosystem management.

Table 5. HUMAN ENERGY EXPENDITURES IN 1963 (from United Nations data; Harper, 1966).

Energy source	Energy used
	$(10^{15} kcal)$
Human labor	0.3
Work animals	1.3
Fuel wood, manure, waste	4.5
Coal lignite	13.1
Petroleum	9.7
Gas	4.8
Water	0.6
TOTAL	34.3

Export

Another form of work of ecosystems is the export of energy outside of the system. Usually export is accomplished through the agencies of wind or water movements, the ecosystem itself being relatively passive. Nevertheless, export may be an extremely important process, especially in riverine systems. Here the biota may depend upon export from terrestrial systems since the rate of water movement or turbidity of the water prevents much photosynthesis from occurring in the river itself. An example of this type of phenomena is the river Thames (Mann, 1964). The requirement of the fish in the river was 18 times the standing crop of available bottom organisms. Even though the bottom fauna have a rapid turnover, 40 percent of the intake for one of the two most important species of fish came from surface drift of terrestrial insects. Further, the bottom organisms feed on organic matter from trees fringing the river. Root Spring (Teal, 1957) is another example, but here the dependency on import of organic matter is even greater. About 77 percent of the energy input to this system was from import of debris and insect immigration.

The Energy Flux to Man

Finally, we might consider the energy flux through the populations of the dominant animal on the planet—*Homo sapiens*. The world estimates of energy use by man are given in Table 5. It is obvious that our present human culture depends upon energy sources derived from photosynthesis and storage in geological time. Man uses about 6.1 x 10^{15} kcal directly from the annual world net production of 655 x 10^{15} kcal, or about 1 percent, but he uses an equivalent of about 5 percent in combustion of fossil fuel. If this energy is used in maintenance of the biosphere, including those artificial systems developed by man, this will produce a quite different effect than if the energy is used to disorder further the biosphere. Unlike other organisms, man has nonrenewable energy resources available to him. If these are used to discover and realize the proper relationship between artificial systems, equilibrium systems, and systems undergoing repair on the planet, mankind could anticipate maintenance of environmental quality and sustained high productivity of the biosphere.

Literature Cited

Andrzejewska, L. 1967. Estimation of the effects of feeding of the sucking insect *Cicadella viridis* L. (Homoptera-Auchenorrhyncha) on plants. *In* K. Petrusewicz [ed.] *Secondary Productivity of Terrestrial Ecosystems*, pp. 791-805. Polish Acad. Sci., Warsaw.

Bolin, B., and C. D. Keeling. 1963. Large-scale atmospheric mixing as deduced from the seasonal and meridional variations of carbon dioxide. J. Geophysical Res., *68*:3899-3920.

Bonner, J. 1962. The upper limit of crop yield. Science, *137*:11-15.

Bowen, H. J. M. 1966. *Trace Elements in Biochemistry*. Academic Press, New York.

Bray, J. R. 1964. Primary consumption in three forest canopies. Ecology, *45*:165-167.

Chevalier, R. 1948. Biogíographic et écologie de la forêt dense ombrophile de la Côte d'Ivoire. Rev. Bot. Appl., *28*:101-115.

Clarke, G. L. 1946. Dynamics of production in a marine area. Ecol. Monogr., *16*: 321-335.

Cummins, K. W. 1967. Calorific equivalents for studies in ecological energetics. Mimeographed Univ. Pittsburgh. 52 pp.

Duever, M., and J. Ewel. 1971. A structural description of second growth vegetation in eastern Panama and northwestern Colombia. *In* F. B. Golley [ed.] *Mineral Cycling in Tropical Forest Ecosystems*. Univ. Georgia Press, Athens, in manuscript.

Fogg, G. E. 1968. *Photosynthesis*. Amer. Elsevier Publ. Co., New York.

Gates, D. M. 1962. *Energy Exchange in the Biosphere*. Harper and Row, New York.

Giddens, J. 1957. Rate of loss of carbon from Georgia soils. Soil Sci. Soc. Amer. Proc., *21*:513-515.

Golley, F. B. 1961. Energy values of ecological materials. Ecology, *42*:581-584.

Golley, F. B. 1969. Caloric value of wet tropical forest vegetation. Ecology, *50*: 517-519.

Harper, R. A. 1966. The geography of world energy consumption. Geography, *65*:302-315.

Holeman, J. N. 1968. The sediment yield of major rivers of the world. Water Resources Res., *4*:737-747.

Hutchinson, G. E. 1954. The biochemistry of the terrestrial atmosphere. *In* G. P. Kuiper [ed.]. *The Earth as a Planet*, pp. 371-433. Univ. of Chicago Press, Chicago.

Judson, S., and D. F. Ritter. 1964. Rates of regional denudation in the United States. J. Geophysical Res., *69*:3395-3401.

Kaczmarek, W. 1967. Elements of organization in the energy flow of forest ecosystems (preliminary notes). *In* K. Petrusewicz [ed.] *Secondary Productivity of Terrestrial Ecosystems,* pp. 663-678, Polish Acad. Sci., Warsaw.

Kira, T., and T. Shidei. 1967. Primary production and turnover of organic matter in different forest ecosystems of the western Pacific. Jap. J. Ecol., *17:*70-87.

Langbein, W. B., and S. A. Schumm. 1958. Yield of sediment in relation to mean annual precipitation. Trans. Amer. Geophysical Union *39:*1076-1084.

Lieth, H. 1962. Die Stoffproduktion der pflanzendecke. Gustav Fischer Verlag, Stuttgart.

Lieth, H. 1964. Versuch einer kartographischen Darstellung der Produktivitat der pfanzendecke auf der Erde. Geogr. Taschenbuch. Franz Steiner Verlag, Wiesbaden, pp. 72-80.

Linton, D. L. 1965. The geography of energy. Geography, *50:*197-228.

Long, F. L. 1934. Application of calorimetric methods to ecological research. Plant Physiol., *9:*323-337.

Lotka, A. J. 1924. *Elements of Physical Biology.* Reprinted 1956, Dover Pub., New York.

Mann, K. H. 1964. The pattern of energy flow in the fish and invertebrate fauna of the River Thames. Vehr. Internat. Verein. Limnol., *15:*485-495.

Mason, B. 1966. *Principles of Geochemistry.* (3rd ed.). Wiley and Sons, Inc., New York.

Menhinick, E. F. 1967. Structure, stability, and energy flow in plants and arthropods in a Sericea lespedeza stand. Ecol. Monogr., *37:*255-272.

Misra, R. 1968. Energy transfer along terrestrial food chain. Tropical Ecol., *9:* 105-118.

Misra, R., J. S. Singh, and K. P. Singh. 1968. A new hypothesis to account for the opposite trophic-biomass structure on land and in water. Curr. Sci., *37:* 382-383.

Odum, E. P., C. E. Connell, and L. B. Davenport. 1962. Population energy flow of three primary consumer components of old-field ecosystems. Ecology, *43:* 88-96.

Odum, H. T. 1956. Efficiencies, size of organisms, and community structure. Ecology, *37:*592-597.

Odum, H. T. 1957. Trophic structure and productivity of Silver Springs, Florida. Ecol. Monogr., *27:*55-112.

Odum, H. T. 1968. Work circuits and system stress. *In* H. E. Young [ed.] *Symposium on Primary Productivity and Mineral Cycling in Natural Ecosystems,* pp. 81-138. Univ. of Maine Press, Orono.

Odum, H. T., and J. Ruíz-Reyes. 1970. Holes in leaves and the grazing control mechanism. *In* H. T. Odum [ed.] *A Tropical Rain Forest,* pp. I-69-I-80. Div. Tech. Info., USAEC, TID-24270 (PrNC-138).

Ovington, J. D. 1965. Organic production, turnover and mineral cycling in woodlands. Biol. Rev., *40:*295-336.

Ovington, J. D., and D. Heitkamp. 1960. The accumulation of energy in forest plantations in Britain. J. Ecol., *48:*639-646.

Reichle, D. E., and D. A. Crossley, Jr. 1967. Investigation on heterotrophic productivity in forest insect communities. *In* K. Petrusewicz [ed.] Secondary Productivity of Terrestrial Ecosystems, pp. 563-587. Polish Acad. Sci., Warsaw.

Richards, D. W. 1964. *The Tropical Rain Forest, an Ecological Study.* Cambridge Univ. Press, Cambridge.

Rutten, M. G. 1966. Geologic data on atmospheric history. Palaeogeography. Palaeoclimat. Palaeoecol., 2:47-57.

Schmitt, W. R. 1965. The planetary food potential. Ann. N. Y. Acad. Sci., 118: 647-718.

Stamp, L. D. 1960. Our Developing World. Faber and Faber, London.

Tadaki, Y., N. Ogata, and Y. Nagatomo. 1965. The dry matter productivity in several stands of Cryptomeria japonica in Kyushu. Bull. Govt. Forest. Expt. Stat., 173:46-66.

Teal, J. M. 1957. Community metabolism in a temperate cold spring. Ecol. Monogr., 27:283-302.

Tilly, L. J. 1968. The structure and dynamics of Cone Spring. Ecol. Monogr., 38:169-197.

Trojan, P. 1967. Investigations on production of cultivated fields. In K. Petrusewicz [ed.] Secondary Productivity of Terrestrial Ecosystems, pp. 545-561. Polish Acad. Sci., Warsaw.

Westlake, D. F. 1963. Comparisons of plant productivity. Biol. Rev., 38:385-426.

Whittaker, R. H. 1970. Communities and Ecosystems. Macmillan Co., New York.

Wiegert, R. G., and F. C. Evans. 1967. Investigations of secondary productivity in grasslands. In K. Petrusewicz [ed.] Secondary Productivity of Terrestrial Ecosystems, pp. 499-518. Polish Acad. Sci., Warsaw.

Discussion

QUESTION: Dr. Golley, you were speaking about using the nonrenewable fuels to support a maintenance system. But what happens when those nonrenewable fuels run out or become scarce?

DR. GOLLEY: I would argue that we should use the stored energy available to us to discover as much as we can about the system to prepare for that time. In other words, we should build upon present energy stores to develop new energy sources. If we are successful, then we will no longer worry about the current nonrenewable fuels but rather, the problems associated with the new fuels. Of course, we may be unsuccessful in developing new sources of power. In that case, we will have to readjust population and societies to fit the available power, and adjustment might be very difficult.

QUESTION: Assuming that earth is a closed system, with continuing burning of fossil fuels, why doesn't the average temperature of the earth just keep going up?

DR. GOLLEY: It would if the earth were a closed system but it is not: the sun continually sends energy to the earth—the input of solar energy and outputs of reflected and re-radiated, etc. energies balance so that the environment at the earth's surface remains relatively constant. I am not especially knowledgable in this field, and I cannot discuss in detail the causes of the temperature changes at the earth surface which have occurred in the past.

Last year I did try to calculate the effect of increasing CO_2 in the atmosphere on the temperature at the earth's surface. The model I've been working with is relatively complicated and, at this time, I can not find sufficiently precise measurements in the literature to make it work. This is all rather disturbing because we have had some ecologists speak recently in a very positive manner about the effect of increasing CO_2. I suggest that you look into the data yourself if you are concerned with this problem.

QUESTION: How do we effectively gain control of the biosphere so we can implement some sound ecological model? And is it really possible to implement any types of activity while maintaining environmental quality under a system where decision making is as nebulose as ours?

DR. GOLLEY: I will disagree with your proposition that we should "gain control of the biosphere." Rather, I suggest that we should develop a strategy so that we can continue to live as part of the biosphere. The disagreement is not semantic. We are a part of cybernetic systems (ecosystems) just like any other living population. Since, by definition, cybernetic systems operate by error control, we can expect that we will err in our interactions with other parts of the biosphere. If this is true, then it is illogical, indeed it is suicidal, to think that we can bring the biosphere under our control and provide the management to maintain biosphere stability. Man has always relied on the nonhuman parts of the biosphere to repair mistakes in ecosystem management and to provide control of human numbers. If we wish to live in a biosphere similar to that present today or present in the past, we must limit our control of the biosphere, so that sufficient unmanaged habitats and areas exist to provide the plants, animals, and microorganisms necessary in repair processes. We must preserve the repair processes to bring the system back to an equilibrium and permit us to start over again. I would argue that this is one of the best reasons for the preservation of what we call "wilderness." It is essential for our survival because we need something to repair the inevitable errors that we are going to make. In other words, we must accept our position as a part of the biosphere, subject to biosphere processes, and not assume dominance even if we can.

The second part of your question concerns the decision-making process. Decisions for ecosystem management can be made by an ecological autocrat or by all of us collectively. In either case the success of the decision-making depends upon the attitudes underlying the process and the information base available. I believe that our present problems stem from both attitude and lack of adequate information, and not entirely from the way decisions are made. However, it is also my opinion that the best system of decision making is to provide all the people the opportunity to express their choice of the best alternatives for their comfort, well being, and survival. If local, immediate reactions to problems are tempered by an understanding of our role in the biosphere, we have a chance to prosper.

Patterns of Production
in Marine Ecosystems

GORDON A. RILEY

Institute of Oceanography
Dalhousie University, Halifax, Nova Scotia

THE TOTAL RANGE of oceanic productivity is approximately the same as that of the land, as Golley has demonstrated in this volume. Certain shallow water communities, such as coral reefs and turtle grasses (*Thallassia*), are approximately as productive, and for essentially the same reasons, as terrestrial marshlands. At the opposite extreme is the Arctic Ocean with its heavy ice cover and strong vertical stability, which keeps essential nutrients locked out of the surface layer. There, the total annual production is far below the level of one day's growth on a coral reef.

In between, and constituting the major part of the area of the world oceans, is a typically oceanic community in which free-floating phytoplankton in the surface layer are the primary producers, and they support, directly or indirectly, a complicated food web of herbivores, carnivores, and scavengers which extends from surface to bottom.

Regional variations in the primary productivity of this type of community are not very large. The richest areas are three to five times as productive as the poorest ones, excluding special cases, such as the Arctic Ocean. The further elaborations of the food web are more diverse. In areas of more or less comparable primary productivity, we find that in some cases the major food supply is channeled into the production of large carnivores, while in other cases the bacteria and detritus feeders manage to get an inordinate share, so that fish stocks can vary by a factor of ten. A major part of this account will be devoted to reviewing the factual evidence on variations in the patterns of ecosystems and exploring their environmental causes.

The implication in the title, that we are dealing with productivity interrelationships within the ecosystem, is admittedly grandiose. We have not solved the difficult problems of determining rates of production with a satisfactory degree of precision. This can be stated categorically, across the board, all the way from primary production to the largest elements in the food web. The account could be cut to half its present length if we had the kind of unequivocal information that is really needed. A great deal of time must be spent in qualifying statements and speculatory discussions about the real meaning of the observations, which are mainly in terms of field data on biomass.

To pin the subject down in a somewhat specific way, I shall compare and contrast three widely differing marine communities: 1) the northwestern Sargasso Sea, a deepwater, impoverished community; 2) a tropical upwelling area off the west coast of Africa, which I believe to be one of the richest oceanic areas in the world; 3) a shallow water and somewhat estuarine situation in temperate waters, namely Long Island Sound, which my colleagues and I have studied rather intensively. Long Island Sound is an aberrant situation, and to broaden the picture, it will be compared with adjacent coastal areas which have not been so severely altered by effluents from the land.

My discussion is largely limited to marine ecosystems, the only aspect of the subject that I feel I can deal with competently. There is no precise parallel between these open ocean systems and terrestrial ecosystems. They are most nearly analogous to grassland communities. Both are of intermediate productivity. In both cases, the plant community is small in terms of total biomass but is produced rapidly and is kept to a low level by constant cropping. Both support a diversified system of herbivores and carnivores. The grassland produces a residuum of humus which is worked over by soil fauna and flora. The marine equivalent is a residuum of particulate and dissolved organic matter in the water column and in bottom sediments. Its organic content is two orders of magnitude larger than that of the living biomass, and its accumulation obviously is due to its refractory nature. Yet, in time, it does get utilized, and contributes to ecosystem productivity, as evidenced by the relatively small organic content that is left in deep sea sediments.

If one looks for other parellels between marine and terrestrial situations, the possibilities are more constrained. It is easy enough to extrapolate the limited areas of coral reefs, turtle grass, and eel grass (*Zostera*) to terrestrial bog and marsh situations. However, there is nothing in the marine environment remotely resembling a forest situation. The giant kelp beds, as well as less impressive growths of inshore attached algae, have an equally high rate of primary production, and they support a diversified fauna and flora, but this community is much less complex than that of the forest. Algal fronds grow and decompose readily, and there is recent evidence that the particulate and dissolved organic materials which they release can provide a significant amount of sustenance for the planktonic community, but there is no great accumulation of woody tissue nor enough physical structure in the community to provide the sort of diversity in niches that one finds in a forest.

Sargasso Sea

The Sargasso Sea has been studied more intensely than most other subtropical regions, but it probably typifies vast reaches of blue waters lying within the subtropical gyres that are present within all the major basins.

In the Sargasso Sea off Bermuda, there is an annual seasonal temperature range of about 18 - 27° C in the surface waters. A summer thermocline develops at a depth of 50 m or a little more; in winter, the water is mixed to a depth of about 150 m. At that time, nutrients presumably are released to the surface by vertical mixing; however, this is not clearly evident in the data. There is enough light to permit a fair amount of phytoplankton growth through the winter. The net result is that the quantity of phytoplankton is slightly larger in winter than in summer, and the concentration of nutrients in winter is not markedly greater than the impoverished summer level. This situation has been described by Riley (1957) and Menzel and Ryther (1960). There is also a small, brief diatom flowering in March or April when thermal stability is becoming established. This suggests that the combination of low light intensity and vertical turbulence is somewhat limiting prior to this time; on the other hand, the supply of nutrients is insufficient for more than a faint shadow of the kind of flowering that occurs in higher latitudes.

The concept that these blue waters are a marine desert is not entirely correct. It is true that the concentration of phytoplankton is only a tenth or even as little as 1/30 of the quantity in temperate waters, but it is a deeply distributed population, which is also able to photosynthesize effectively down to a depth of 80 m or more in these transparent waters. As a result, the total population per unit of sea surface and its total productivity are of the order of a third that of productive temperate regions rather than a tenth or less.

The average productivity in terms of C^{14} uptake as observed in a two-year series of observations obtained by Menzel and Ryther (1960) off Bermuda was about 200 mg $C/m^2/day$. This value was obtained in a series of simulated *in situ* experiments by standard methods in practice at that time, and the true level of productivity is somewhat in doubt in the light of more recent work. Goldman (1968) examined the errors due to self absorption of C^{14} by phytoplankton on filters and found that in extreme cases the values obtained by the routine method could be as little as 20 percent of true productivity. A more conservative estimate of error is given by the recommendation of the International C^{14} Laboratory that the older values be increased by 45 percent to allow for self absorption. In addition, there is extracellular release of dissolved organic substance by phytoplankton. Hellebust (1965) and others have examined this problem in experimental cultures, and Choi (1970) analyzed natural populations in subtropical waters of the Gulf Stream and northern Sargasso Sea. He reported that extracellular release was of the order of 25 percent of total production. In short, total productivity in the Bermuda region could be nearly twice the value reported by Menzel and Ryther (1960).

Another paper by the same authors (Menzel and Ryther, 1961) described a two-year series of zooplankton collections in the same area and experiments on their oxygen consumption to estimate minimum food requirements. The total quantity of zooplankton in the upper 500 m averaged 1.08 g dry weight, and the food needed to satisfy their respiratory requirements was 135 mgC/m²/day.

These waters also contain considerable quantities of nonliving particulate organic matter which is inhabited by an assortment of bacteria, protozoa, and small algae, the latter presumably living heterotrophically in waters below the euphotic zone. The total quantity of particulate carbon, living and dead, is of the order of 10-140 $\mu gC/1$ in the surface layer and declines to a low and relatively constant level of about 18 $\mu gC/1$ in waters below about 900 m. The food supply of the associated population of microorganisms presumably includes both the particulate matter itself and dissolved organic matter in the water. It is a reasonable assumption that most of the extracellular release by phytoplankton is used up by such organisms in the upper few hundred meters, for Pocklington (1970) has shown that the composition of free amino acids in sea water changes with depth, becoming more limited in deep water and more nearly typical of what one would expect as a result of animal excretion rather than plant release.

The quantity of organic matter required by the so-called ultraplankton in the Sargasso Sea has not been determined. Pomeroy and Johannes (1968) have examined this question in some other areas and have reported a requirement approximately equal to "uncorrected" C^{14} uptake. Possibly, then, the ultraplankton obtains a slightly larger proportion of the phytoplankton production than does the herbivorous zooplankton.

Table 1. BALANCE SHEET OF ESTIMATED PRODUCTION AND CONSUMPTION IN THE
SARGASSO SEA OFF BERMUDA

Process	Depth	Component	Process rate
	m		$mgC/m^2/day$
Production	0–100..........................uncorrected C^{14} uptake		200
		estimated total production	350–400
Consumption	0–300..........................zooplankton		135
		ultraplankton	200
	300–900.......................zooplankton + ultraplankton		40
	900–bottom..................zooplankton + ultraplankton		6
	On bottom..................benthic fauna and bacteria		4

The remainder of the food web in the surface layer is not well known, but certainly it is not abundant. There is a meager population of flying fish (Exocoetidae) and other zooplankton feeders, slightly augmented at night by myctophids and other fishes coming up from middepths. Larger carnivores such as tunas and sharks are caught occasionally, but never in abundance except for rare schools of tuna, which appear merely to be migrating through the area.

In deeper waters, Menzel and Ryther (1961) reported that the total quantity of zooplankton below 500 m is approximately twice that of the upper 500 m. However, there is little doubt that low temperature and near starvation conditions lead to a low metabolic rate and rate of production in most of this fauna. Riley et al. (1965) analyzed various kinds of data bearing on this subject and concluded that the total carbon consumption below 300 m probably is not more than about 50 mg/m²/day. Moreover, the greater part of this was consumed at middepths, and the remaining small fraction was about equally divided between the sparsely inhabited bathypelagic zone and the bottom.

At this point, it is possible to draw up, in a very tentative way, a balance sheet of production and consumption in a series of depth ranges from surface to bottom. It will be apparent from the foregoing that this is little more than guesswork, but for what it may be worth, the results are shown in Table 1.

The bathypelagic and bottom situations require some further comments, although there is little solid information on the subject. The bathypelagic assemblage of animals, ranging from copepods to fishes, is probably very largely carnivorous and necrophagous in habit and depends upon a slow biological diffusion of food into deep water

(Riley, 1951; Vinogradov, 1962). Some of the copepods have folia-
ceous appendages, suggesting a filter feeding habit. Their mode of support is problematical. Riley (1970) has examined food requirements
of surface zooplankton living in near-starvation conditions in cold
winter waters, where a minimum sustaining diet is about 17 μgC/1.
This is essentially the amount of total particulate organic carbon available in deep water; however, Gordon (1970) found that only about
20 percent of these deepwater collections could be hydrolyzed in laboratory experiments with trypsin, chymotrypsin, and amylase. We are
left with the indefinite alternatives that the deepwater filter feeding
fauna might have more effective digestive enzymes than the ones used
in these experiments, or they might have a very depressed metabolic
requirement, or they might supplement filter feeding by selective capture of larger particles.

Bacteria and heterotrophic algae have been seen in bathypelagic
collections from this region. Quantitative estimates of biomass are
lacking, although there is no reason to doubt that here, as in other
areas that have been studied more carefully, the organic content of
heterotrophs is a very small fraction of the total. ATP analyses (Hamilton, et al., 1968) in other areas suggest that the biomass of ultraplankton is of the order of one percent of the particulate organic matter. They are able to utilize low concentrations of organic substrates of
the order of 10^{-7} to 10^{-8} M, such as are found in these waters (Parsons
and Strickland, 1962), but the daily rate of growth at these concentrations appears to be only 10 percent or less of the standing crop, which
is small compared with surface forms.

Estimates of the sinking rate of particulate organic matter indicate that the rate of addition to the deep ocean bottom is of the order of
5 mgC/m²/day (Riley, 1970). The particles in the water column have
an organic content that averages about 20 to 30 percent of the total
dry weight. The bottom sediments themselves generally have an organic
content of less than one percent. This implies extensive utilization of
organic matter on the bottom. Compaction of the particles on the bottom and imprisonment of interstitial water logically can lead to a
richer environment for heterotrophs and a better opportunity for the
eventual dissolution of relatively refractory materials. The bottom
fauna that develops is small compared with shallower and more productive waters (Sanders et al., 1965), but it compares favorably with
the large but sparsely populated bathypelagic zone lying above.

The Eastern Tropical Atlantic

The Benguela Current sweeps up the west coast of South Africa toward the Equator. Under the influence of the Southeast Trades, surface water is transported away from the coast. The thermocline becomes shallow and sharp, and removal of the surface layer leads to replacement by cool and nutrient-rich water upwelling from the upper middepths.

Farther north, the diffuse Azores Current flows down around the northern bulge of the African continent and intensifies as it flows into the Gulf of Guinea. Here, too, as a result of wind stress from the Northeast Trades, there is offshore movement and upwelling.

A similar phenomenon occurs along the west coast of South America. Divergence along the Equator in both oceans extends the upwelling zone into midocean. In the Indian Ocean, monsoon winds produce intense local and seasonal upwelling. These tropical areas of upwelling possibly are the most productive regions in all of the world oceans, although in the present state of our knowledge, any such statement is debatable.

Expeditions by the *Meteor* and *Dana* in the mid to late twenties explored the general situation off the west coast of Africa. Upwelling was discontinuous and streaky. Newly upwelled water is cool and nutrient rich, but poor in plankton. However, phytoplankton blooms come along quickly and are followed by heavy zooplankton production. The spotty nature of the phenomenon, combined with a lack of good year-around data, even up to the present time, makes it difficult to determine either the average rate of upwelling or the productivity of the region as a whole.

Early expeditions established the fact that the middepth oxygen minimum layer and associated phosphate and nitrate maxima were more strongly developed here than in any other part of the Atlantic Ocean, suggesting a rich middepth fauna and flora. Minimum oxygen concentrations were of the order of 0.6 ml/l, as contrasted with values of the order of 2.5 ml/l in the centers of the subtropical gyres. The report by Jesperson (1935) on *Dana* collections of macrozooplankton was also indicative of the richness of this area.

A later Danish expedition, the *Galathea*, skirted the West African coast in the first leg of its circumnavigation of the globe. This expedition marked the advent of the C^{14} method for measuring carbon fixation by phytoplankton. There were several papers describing the results, the most definitive being that of Steemann-Nielsen and Jensen (1957). Individual values varied widely, a large proportion falling

Table 2. DAILY PRIMARY PRODUCTION AND ZOOPLANKTON BIOMASS AT THREE OCEANIC LOCATIONS (from Greze, *et al.*, 1969)

Depth	Location		
	5°N-4°S 15°W	1.5°N-1.5°S 35°W	15°S 17-35°W
	mgC/m²	*mgC/m²*	*mgC/m²*
0-100 m............Primary production	880	480	252
Herbivores	397	658	140
Carnivores	377	572	148
Total zooplankton	774	1,230	288
0-500 m............Total zooplankton	1,750	1,900	680

within the general range of 0.5-1.0 gC/m²/day, but ranging up to 2.4 gC/m²/day in Walvis Bay in Southwest Africa. Walvis Bay is well known from other work as an area where nutrient enrichment can sometimes be excessive, to the point where the ecosystem goes awry. Decomposition of phytoplankton in the deeper water exhausts the oxygen supply, leading to H_2S production and fish kills. Similar occurrences have been noted off the west coast of Peru, where the term "Callao painter" refers to the fact that H_2S production in the harbor of Callao blackens boat paints. These are extreme cases in which natural fertilization can approximate the harmful effects of man-made pollution, but in general, the ecosystem remains in balance in these upwelling areas, and there is enhanced production in all parts of the food web.

Corcoran and Mahnken (1969) and Greze *et al.* (1969) have provided more extensive data on C^{14} fixation. There is some seasonal variation due to a seasonal shift of the climatic equator, but in general high values of 1 gC/m²/day are common in the richest part of the Gulf of Guinea and occur more patchily in other parts of the immediate coastal area. Productivity declines seaward, but remains moderately high for some distance westward in the equatorial divergence. In general, one can say that probably the richest areas have a production rate varying around five times that of uncorrected C^{14} uptake in the Sargasso Sea, and rather large areas extending to the westward have a value of the order of two to three times that of the so-called subtropical deserts.

Greze *et al.* (1969) measured zooplankton biomass as well as C^{14} fixation, and they further attempted to divide the zooplankton into herbivorous and carnivorous fractions. A portion of their data is shown in Table 2. The values given are averages for stations in the richest area that they examined, and in two poorer areas to the south and southwest.

The poorest area examined is not very different from the Sargasso Sea station described earlier. Other areas range up to roughly three times that level, and they probably are too far out to sea to represent peak levels of production.

The results are remarkable in that the biomass of primary carnivores appears to be nearly the same as that of herbivores. The authors admit some difficulty in trying to evaluate food habits of some of the organisms, but the results may be valid. By analogy, there are well-known instances in which a highly productive phytoplankton population maintains a large zooplankton assemblage but is kept to a low level of biomass by heavy cropping. In the present case, an inordinately large population of carnivores conceivably could maintain a predation pressure that would keep the herbivores down to a low biomass with a high rate of turnover.

Voss (1969) described tows made with midwater trawls in a series of strata from surface to near bottom. The catches were mainly carnivores of medium size, and represented intermediate levels in the food heirarchy. Without going into the interesting details of population structure, it is sufficient here to say that his catches had about ten times the biomass obtained in similar kinds of tows in the Carribbean and western tropical Atlantic.

As to small particulate matter in the water column, Wangersky and Gordon (1965) found concentrations of about 55-200 μgC/1 in the surface layer. Deepwater values averaged 48 μgC/1 in stations that were fairly close inshore, 35 μg a little farther from the coast, and declined to 18 μg/1 in subtropical waters to the northwest. The data leave no doubt that the whole ecosystem, from surface to bottom, benefitted from the high level of surface productivity. Zei (1969) reported that there is an abundance of sardines and related species, as might be expected in waters where there is an abundant zooplankton population. The pelagic fish provide the basis for a local canoe fishery in inshore waters, and probably could support a modern fishery of much larger proportions, which is now beginning to develop. They also supply food for tunas and other large carnivores.

Wise and LeGuen (1969) analyzed the Japanese long-line fishery from its early stages in the midfifties to 1963. The total catch during this period was some 12 million fish. Four species of tunas together constituted over 90 percent of the catch, the remainder being marlins, swordfish, and a few other species. The catch declined during this period, and the stock apparently was beginning to be overfished at a very early stage in the development of the fishery. These authors arbitrarily divided the Atlantic Ocean between 30°N and 30°S into ten areas for purposes of analyzing catch statistics. The catch per unit of effort was greatest in the Gulf of Guinea, but in general the regional variations in catches of large pelagic fish were not as marked as variations in primary productivity. This was ascribed to the fact that many of these fish perform extensive migrations, so that the quantity found in a particular location at any one time is not truly indicative of the capacity of the area to support growth.

Information about invertebrate bottom fauna is scanty, but according to Longhurst (1969), the benthic fish population is a fairly rich one and much more diversified than is common in many temperate waters. In surveys off Sierra Leone and Nigeria, 15 species or more were present in significant numbers, and no one species constituted more than 15 percent of the total. The total catch in the richest area, between 10° and 13°N, was 11 g wet weight/m², and off the Congo at 3-6°S it was about 8 g/m². There were some poorer areas with minimum populations of the order of 2 g/m².

Long Island Sound

Long Island Sound is a shallow and semi-enclosed basin with slightly reduced salinity and is, in general, a somewhat estuarine environment of moderately high productivity but limited species diversity, as is common in such areas. There is a pronounced seasonal cycle of phytoplankton, with a midwinter minimum and a large diatom flowering in late winter or early spring. In late spring and summer, the phytoplankton crop is at a lower and more nearly constant level. However, there is enough tidal mixing so that the water column is only slightly stable, facilitating vertical transport of nutrients regenerated from the bottom and maintaining productivity at a higher rate than one would expect to find in more stable waters offshore. Autumn flowerings are small and sometimes absent.

The water is turbid. The euphotic zone rarely extends below 10 m and may be considerably less. The rate of growth is rapid near the surface, but tidal stirring tends to mix the phytoplankton into deeper waters where growth is ineffective. The net result is a large standing stock with a relatively low overall rate of turnover.

Estimates of phytoplankton production include light and dark bottle oxygen experiments by S. Conover (1956) and physical oceanographic computations by Riley (1956) based on observed oxygen distribution and estimated vertical mixing rates. These are not quite comparable to C^{14} experiments, precluding precise comparison. The estimate for mean annual photosynthesis as determined by light and dark bottle experiments is equivalent to 1,300 mgC/m²/day. Physical oceanographic estimates of net production averaged 530 mgC/day. The latter probably more or less corresponds to C^{14} uptake corrected for self-absorption. Thus, the productivity of Long Island Sound appears to be intermediate between the other two areas that have been discussed.

The zooplankton population is volumetrically abundant, but limited as to species content (Deevey, 1956), as is common in such areas. The total crop in the vertical column, which averages only 20 m deep, is about 2 g/m^2 (dry weight). This is nearly twice as much as the more thinly extended population in the upper 500 m of the Sargasso Sea, but less than the total population, surface to bottom, in the latter area.

Clearly, the Long Island Sound zooplankton does not get as large a share of total primary production as the populations in some other areas, and there are several reasons why this might be so. R. Conover (1956) reported that the dominant copepods are less efficient in their feeding than some open-ocean forms which apparently are largely excluded from the Sound by low salinity. On the basis of his feeding experiments, it is estimated that the zooplankton consumes only about a quarter of the primary production.

In an area of this sort, there is always a significant quantity of phytoplankton in the vicinity of the bottom. The zooplankton is in direct competition with benthic filter feeders, and some of the plankton settles on the bottom and is available to other portions of the benthic community.

The concentration of nonliving particulate organic matter in the water column generally exceeds the biomass of living phytoplankton (Riley, 1959), and it harbors an abundant population of bacteria (Riley and Altschuler, 1967) and other ultraplankton. Consumption by these forms appears to be larger than that of zooplankton (Riley, 1956), and there is no good explanation as to why the latter does not utilize these minor sources of food supply more effectively. Possibly when the concentration of food is very large, there are density-dependent factors other than food supply which limit zooplankton growth.

Sanders (1956) examined the invertebrate bottom fauna and estimated its annual productivity, and Carey (1967) determined food requirements in terms of oxygen consumption of major components of the fauna and total oxygen consumption in sediment cores. The quantity of animals obtained by ordinary dredging and sieving techniques was moderately large, but Carey's work showed that their respiration was at most 17 percent of total oxygen consumption on muddy bottoms, which constituted some 80 percent of the area of Long Island Sound. The remainder of the respiratory requirement could be ascribed to bacteria, nematodes, and other small organisms which es-

Table 3. MEAN ANNUAL STANDING CROPS AND MEAN DAILY PRODUCTION AND
CONSUMPTION OF ORGANISMS IN LONG ISLAND SOUND

	Standing crop	Production	Consumption
	gC/m^2	$mgC/m^2/day$	$mgC/m^2/day$
Phytoplankton	8	530 (net)	
Zooplankton	1	27	140
Other pelagic organisms	----	----	230
Benthic invertebrates	5	36 (14)	45
Benthic microbenthos	----	----⎰	130
Benthic fishes	0.2	0.5⎱	

caped the sieve collections. Weiser and Kanwisher (1961) have described a somewhat similar situation in an intertidal mudflat.

The benthic fish population (Richards, 1963a) was less than 1 g/m² (wet weight). This is probably of the order of 1/15 the quantity in the richest parts of the African coast. Reasons for this scarcity are obvious. The abundant microfauna and flora on these soft bottoms were of little use as fish food, nor were the molluscs, which constituted a significant fraction of the larger fauna. The molluscs supported a population of starfish that was considerably larger than that of fin fish. Combining these two under the general category of top carnivores in the system, they add up to perhaps half the tropical African fish biomass, but as a practical commercial fishery, Long Island Sound is distinctly marginal.

This general situation, stemming from the soft character of the bottom, is basically due to the fact that Long Island Sound acts as a catchment basin for silts and clays from several large rivers. More open coastal waters to the eastward have essentially the same level of primary productivity (Riley, 1955). However, the bottom is mainly sand, or mixed sand and mud, and the invertebrate benthic fauna is dominated by crustacea, which are excellent fish food. Thus, it is not surprising that Merriman and Warfel (1948) reported a moderately large benthic fish population of about 9 g/m².

Riley (1956) summarized the information that was available at that time on the biomass of various groups of organisms in Long Island Sound and developed a general balance sheet of production of carbon by phytoplankton and its respiratory utilization by consumers. These data are presented in Table 3, in somewhat modified form and with addition of other information obtained subsequently. Figures on phytoplankton production, and consumption by the remainder of the population, are largely derived from physical oceanographic computations of net oxygen changes, and basically there are two estimates, one for total consumption in the water column, and the other for consumption on and in the bottom. However, these are supplemented by experimental information. The work by R. Conover (1956) permitted a generalized estimate of consumption by zooplankton. This value is shown in the table, and the difference between total water column respiration and the zooplankton estimate is designated "other pelagic organisms." Logically, this includes both ultraplankton and large animals that are not adequately sampled by plankton nets, but the contribution by large forms probably is minor. Similarly, data from Carey (1967) are used to estimate the respiratory requirement of macrobenthos, leaving a residual amount that is allocated to "other benthic organisms."

The figure for zooplankton production is a crude order-of-magnitude estimate. The copepods which constitute the dominant species have a generation time of about a month or five weeks and breed more or less throughout the year, although the intensity varies seasonally. The figure in the table assumes ten crops per year, which is probably a fairly reasonable estimate.

Estimates of production by benthic invertebrates and fishes were obtained by Sanders (1956) and Richards (1963b) respectively, using life history data for the species in question which were compiled in a much more careful way than the zooplankton estimate. Sanders' figure of 36 mgC/m²/day is an arithmetic average of all of his stations. The smaller number in parentheses is his estimate for the soft bottom station where Carey did his later work on respiratory requirements.

The sum of production and consumption is presumed to equal total assimilation. Thus, the computed production efficiency in terms of food assimilated (Ecological Efficiency 11 of Kozlovsky [1968]: [net productivity at n] $\div$ [assimilation at n]) is 16 percent for zooplankton and 24 percent for benthic invertebrates. The latter figure probably is too high, but there are too many uncertainties in this kind of analysis to pinpoint the source of error.

No attempt has been made to estimate the productivity of the ultraplankton or microbenthos. The efficiency could be in excess of 20 percent, although this remains to be established. Small and rapidly growing organisms tend to be efficient in their growth, and in addition, both of these groups have access to dissolved amino acids and other compounds of low molecular weight which can be converted into cell substance at minimal biochemical cost.

Conclusions

It will be apparent that this account has dealt more with biomass patterns than productivity patterns in ecosystems. As indicated earlier, we still know very little about marine productivity. There are several different ways of measuring C^{14} fixation, and at least three other techniques of investigating primary production. They measure somewhat different kinds of things, and vary considerably among themselves. Not one can be said unequivocally to conform to our idealized definitions of gross or net primary productivity.

There have been few attempts to measure the productivity of the consumer elements of a whole ecosystem. One of the most complete attempts was that of Harvey (1950) in the English Channel. His paper stands as a monument to the intensive early work of the Plymouth Laboratory in this area, and to the mature oceanographic judgment of the author. Yet, when this has been said, one has to add that some of the early data were of questionable accuracy, and we now know that the ecosystem is much more complicated than it appeared to be 20 years ago. The work in Long Island Sound that was described above was much less complete than Harvey's, and in some respects it is becoming obsolete, too.

Estimates of the productivity of the higher trophic levels have generally depended upon life history studies and the construction of growth and survivorship curves. These take an inordinate amount of time if one tries to deal with any considerable fraction of the ecosystem, and basically they are crude results because we are sampling populations that are very inhomogeneous in space and time.

There have been some excellent laboratory studies on growth in relation to abundance and quality of food, as for example the work of Dickie (1970) on fishes. Fairly successful attempts have been made to assess productivity of laboratory populations in terms of RNA content (Sutcliffe, 1965). There are possibilities for dealing with natural ecosystems with methods less cumbersome than those of the past, but these have not yet quite jelled.

Much of the older literature more or less equated biomass with productivity. There is probably some validity in this, but there are exceptions that we know about and others that we suspect. We accept the concept of a pyramid of production in ecosystems; metabolic losses at each successively higher level are unavoidable. However, the large and long-lived and slow-growing animals in the upper levels of the marine ecosystem may accumulate a larger biomass than the phytoplankton and zooplankton, which produce rapidly and are rapidly removed by predation. Thus, there can be an inverted pyramid of biomass. This is not clearly the case in Long Island Sound, but it is certainly true in

Table 4. RELATIVE RICHNESS OF VARIOUS POPULATIONS OFF THE AFRICAN COAST,
IN LONG ISLAND SOUND, AND IN THE SARGASSO SEA

Populations	Richest		Less rich		Least rich
Primary production	Africa	>	Long Island	>	Sargasso Sea
Zooplankton (surface layer)	Africa	>	Long Island	>	Sargasso Sea
Zooplankton (total)	Africa	>	Sargasso Sea	>	Long Island
Ultraplankton (surface layer)	Long Island	>	Africa	>	Sargasso Sea
Ultraplankton (total) ..	Africa	>	Long Island	>	Sargasso Sea
Pelagic fishes	Africa	>	Long Island	>	Sargasso Sea
Benthic fishes	Africa	>	Long Island	>	Sargasso Sea
Other benthic flora and fauna	Long Island	>	Africa	>	Sargasso Sea

open coastal waters to the east and in the English Channel. It may also be so in west African coastal waters.

At any particular trophic level, one suspects that there may be variations in the ratio of production to biomass, although solid evidence is lacking. Arctic zooplankton tends to breed annually or biennially. In boreal waters, there may be three or four crops in a year. In some temperate and tropical waters, there is almost continuous breeding and a short life span. It is hard to imagine that these different types of life cycles would not affect the general level of productivity. However, in poor subtropical waters, the zooplankton has a low fat storage and a total organic content that is hardly half that of a well-fed population in temperate waters. One suspects that much of their energy is expended to obtain a maintenance ration, and that productivity may be low. If this were so, one would expect the next trophic level to be even more pinched for food. Jesperson (1935) reported pronounced regional variations in macrozooplankton, largely representing carnivorous and omnivorous forms, and the differences were distinctly greater than the known regional variations in small herbivorous zooplankton. There is support for the hypothesis, but it remains as a train of suppositions.

Considering the data at hand, primarily consisting of some imperfect estimates of primary productivity and semiquantitative information on the biomass of consumers, the order of relative richness in the three areas that have been discussed is probably more or less as indicated in Table 4, although, admittedly, there are some guesses in this table.

Primary production, the zooplankton of the surface layer, and the pelagic fishes are closely linked from a trophic point of view. They all have the same order of trophic abundance, and this order depends upon a few factors which determine the level of primary productivity. Overwhelmingly important in this respect is the rate of transfer of nutrients into the surface layer, which depends in turn upon the rate of vertical circulation and the available stock of nutrients in deeper water. The African coastal area is ideal in both respects. Long Island Sound has effective tidal mixing, but lacks an abundant nutrient reserve. There is some slight enrichment due to freshwater drainage and to transport exchange between the Sound and adjacent coastal waters, but the main source is internal biological cycling, which tends to proceed at a slower rate than physical renewal by upwelling. In the Sargasso Sea, there is a good stock of deepwater nutrients, but vertical circulation is weak.

Secondarily, the quantity and degree of seasonal uniformity of radiation are important. In this respect, Long Island Sound is the poorest of the three areas. Less seasonal variation undoubtedly would lead to some increase in primary productivity and a marked increase in zooplankton. Zooplankton growth cannot effectively keep pace with the sudden increase in phytoplankton that occurs as a result of improving radiation in late winter and early spring. A by-product of this imbalance is an excess of unutilized organic detritus, and this appears to be at least partly responsible for the large concentration of ultraplankton in the Sound.

The depth of water significantly affects the general structure of the food web. A long vertical column leads to an essentially pelagic system and often a truncated food web. There is marked development of zooplankton and ultraplankton, so that even in a relatively impoverished system, such as the Sargasso Sea, the total quantity in the column may exceed that of richer, shallow waters, such as Long Island Sound. However, this material is thinly dispersed in the water column, and apparently does not provide an adequate food supply for predators. Quantitative data are inadequate for precise comparisons, but macrozooplankton and pelagic fishes appear to be scanty. Little usable food gets to the bottom, and the benthic fauna and flora are sparse, too, but this is, of course, true of all deep ocean situations.

In contrast, the plankton is concentrated in shallow water regions and is in part available to benthic fauna, as well as to pelagic predators. The large population of pelagic fishes off the African coast has been described. This element of the ecosystem appears to be much less abundant in Long Island Sound, except possibly in summer, when large schools of menhaden (*Brevoortia tyrannus*) move in. Reasons for this are not immediately apparent. Possibly some species are excluded by low salinity, or the copepod fauna, consisting primarily of small species, may be less attractive than that of open coastal waters.

The earlier account has documented the richness of benthic fauna and flora in Long Island Sound and its qualitative deficiencies with respect to support of benthic fishes. Such information on benthic invertebrates off the African coast as is available suggests that they are not remarkably abundant, though better able to supply fish food. However, the methods used were not fully comparable, and any such generalizations ignore the fact that some areas off the African coast have a fairly poor population of benthic fishes. Muddy bottoms off the mouths of the major rivers may have a benthic population structure quite similar to that of the Long Island Sound basin.

All this tends toward the conclusion that the generalities of relative richness of an ecosystem and its component parts are largely explainable in terms of a handful of environmental factors of a mainly physical nature: radiation, vertical circulation, the nutrient pool, depth of water, and the character of bottom sediments. At the secondary level, there is a welter of biological interrelationships which determine the net response of ecosystem structure to the physical environment. Other physical factors, as well as biological ones, presumably are important in determining individual species composition and the degree of species diversity. However, these are somewhat aside from the central themes of ecosystem productivity and biomass, and for these, the simplistic picture that has been painted is probably suitable as a first-order approximation.

In conclusion, some additional comments are in order on similarities and differences between marine and terrestrial ecosystems. Obviously, the chemical limiting factors are different—primarily minor chemical elements in the one case, and water in the other. However, as far as essential elements are concerned, the operation of both types of ecosystems builds a structure and a means of conserving the resources in the form of accretions of unutilized organic matter, and in the case of terrestrial ecosystems, this structure also stores water and helps to conserve that important resource. The marine climate tends to be more equable seasonally, and microclimates are less pronounced. Niche separation due to varying bottom types is very pronounced in intertidal and subtidal waters, and decreases seaward. Only on the bottom is this kind of microhabitat very important, in contrast with the highly structured environment that exists in a forest, for example.

As indicated earlier, the closest simile to the open sea in terrestrial communities is the grassland ecosystem, and perhaps one should specify the savannas of warm climates, where the environment is equable enough to permit some growth on a year-around basis. Variations in soil and availability of water create microhabitats, but this is minor, as in the ocean, and in contrast with more complex terrestrial environments. There is a small biomass of primary producers and a large biomass of long-lived herbivores and carnivores. The parallel can be pushed further, but functionally it begins to break down. There is, for example, a "pelagic" community of insects and birds which is to some degree functionally integrated with land-bound insectivores and small predators, but is not closely allied nutritionally with the larger animals. Also, the large herbivores, by social organization and other adaptations, are much less closely controlled by predation than is the case in the marine community. The latter, then, is simpler and more closely knit along trophic lines and more amenable to analysis in terms of classical prey-predator concepts.

Literature Cited

Carey, A. G. 1967. Energetics of the benthos of Long Island Sound. I. Oxygen utilization of sediment. Bull. Bingham Oceanogr. Coll., *19*:136-144.

Choi, C. I. 1970. Phytoplankton production and extracellular release in Nova Scotia coastal and offshore waters. M.Sc. thesis, Dalhousie University.

Conover, R. J. 1956. Oceanography of Long Island Sound, 1952-1954. VI. Biology of *Acartia clausi* and *A. tonsa*. Bull. Bingham Oceanogr. Coll., *15*:156-233.

Conover, S. A. M. 1956. Oceanography of Long Island Sound, 1952-1954. IV. Phytoplankton. Bull. Bingham Oceanogr. Coll., *15*:62-112.

Corcoran, E. F., and C. V. W. Mahnken. 1969. Productivity of the tropical Atlantic Ocean. Proc. Symp. Oceanography and Fisheries Resources of the Tropical Atlantic (UNESCO Publication):57-67.

Deevey, G. B. 1956. Oceanography of Long Island Sound, 1952-1954. V. Zooplankton. Bull. Bingham Oceanogr. Coll., *15*:113-155.

Dickie, L. M. 1970. Food abundance and availability in relation to production (Introduction). *In* J. H. Steele [ed.] *Marine Food Chains,* pp. 319-324. Oliver and Boyd, Edinburgh.

Goldman, C. R. 1968. The use of absolute activity for eliminating serious errors in the measurement of primary productivity with C^{14}. J. Cons., *32*:172-179.

Gordon, D. C. 1970. Some studies on the distribution and composition of particulate organic carbon in the north Atlantic Ocean. Deep-Sea Res., *17*:233-244.

Greze, V. N., K. T. Gordejeva, and A. A. Shmeleva. 1969. Distribution of zooplankton and biological structure in the tropical Atlantic. Proc. Symp. Oceanography and Fisheries Resources of the Tropical Atlantic (UNESCO Publication):85-90.

Hamilton, R. D., O. Holm-Hansen, and J. D. H. Strickland. 1968. Notes on the occurrence of living microscopic organisms in deep water. Deep-Sea Res., *15*:651-656.

Harvey, H. W. 1950. On the production of living matter in the sea off Plymouth. J. Mar. Biol. Assoc. U. K., 29:97-137.

Hellebust, J. A. 1965. Excretion of some organic compounds by marine phytoplankton. Limnol. Oceanogr., 10:192-206.

Jesperson, P. 1935. Quantitative investigations on the distribution of macroplankton in different oceanic regions. Dana Rep., 7:1-44.

Kozlovsky, D. G. 1968. A critical evaluation of the trophic level concept. Ecology, 49:48-60.

Longhurst, A. R. 1969. Species assemblages in tropical demersal fisheries. Proc. Symp. Oceanography and Fisheries Resources of the Tropical Atlantic (UNESCO Publication):147-168.

Menzel, D. W., and J. H. Ryther. 1960. The annual cycle of primary production in the Sargasso Sea off Bermuda. Deep-Sea Res., 6:351-367.

Menzel, D. W., and J. H. Ryther. 1961. Zooplankton in the Sargasso Sea off Bermuda and its relation to organic production. J. Cons., 26:250-258.

Merriman, D., and H. E. Warfel. 1948. Studies on the marine resources of southern New England. VII. Analysis of a fish population. Bull. Bingham Oceanogr. Coll., 11:131-164.

Parsons, T. R., and J. D. H. Strickland. 1962. On the production of particulate organic carbon by heterotrophic processes in sea water. Deep-Sea Res., 8:211-222.

Pocklington, R. 1970. Dissolved free amino acids of North American coastal waters. Ph.D. thesis, Dalhousie University.

Pomeroy, L. R., and R. E. Johannes. 1968. Occurrence and respiration of ultraplankton in the upper 500 meters of the ocean. Deep-Sea Res., 15:381-391.

Richards, S. W. 1963a. The demersal fish population of Long Island Sound. I. Species composition and relative abundance in two localities, 1956-1957. Bull. Bingham Oceanogr. Coll., 18:5-31.

Richards, S. W. 1963b. The demersal fish population of Long Island Sound. II. Food of the juveniles from a sand-shell locality (Station 1). Bull. Bingham Oceanogr. Coll., 18:32-72.

Riley, G. A. 1951. Oxygen, phosphate, and nitrate in the Atlantic Ocean. Bull. Bingham Oceanogr. Coll., 13:1-126.

Riley, G. A. 1955. Review of the oceanography of Long Island Sound. Deep-Sea Res., 3(suppl.):224-238.

Riley, G. A. 1956. Oceanography of Long Island Sound, 1952-1954. IX. Production and utilization of organic matter. Bull. Bingham Oceanogr. Coll., 15:324-344.

Riley, G. A. 1957. Phytoplankton of the north central Sargasso Sea, 1950-52. Limnol. Oceanogr., 2:252-270.

Riley, G. A. 1959. Note on particulate matter in Long Island Sound. Bull. Bingham Oceanogr. Coll., 17:83-86.

Riley, G. A. 1970. Particulate organic matter in sea water. Advances in Marine Biology, 8:1-118.

Riley, G. A., and S. J. Altschuler. 1967. Microbiological studies in Long Island Sound. Bull. Bingham Oceanogr. Coll., 19:81-88.

Riley, G. A., D. Van Hemet, and P. J. Wangersky. 1965. Organic aggregates in surface and deep waters of the Sargasso Sea. Limnol. Oceanogr., 10:354-363.

Sanders, H. L. 1956. Oceanography of Long Island Sound, 1952-1954. X. The biology of marine bottom communities. Bull. Bingham Oceanogr. Coll., 15:345-414.

Sanders, H. L., R. R. Hessler, and G. R. Hampson. 1965. An introduction to the study of deep-sea benthic faunal assemblages along the Gay Head-Bermuda transect. Deep-Sea Res., *12*:845-867.

Steemann-Nielson, E., and E. A. Jensen. 1957. Primary ocean production: The autotrophic production of organic matter in the oceans. *Galathea* Rep., *1*:49-136.

Sutcliffe, W. H. 1965. Growth estimates from ribonucleic acid content in some small organisms. Limnol. Oceanogr., *10*(suppl.):R253-258.

Vinogradov, M. E. 1962. Feeding of the deep-sea zooplankton. Rapports et Procès-Verbeaux, Réunions du Conseil permanent internationale pour L'Exploration de la Mer., *153*:114-120.

Voss, G. L. 1969. The pelagic midwater fauna of the eastern tropical Atlantic with special reference to the Gulf of Guinea. Proc. Symp. Oceanogr. and Fisheries Resources of the Tropical Atlantic (UNESCO Publication):91-99.

Wangersky, P. J., and D. C. Gordon. 1965. Particulate carbonate, organic carbon, and Mn^{++} in the open ocean. Limnol. Oceanogr., *10*:544-550.

Weiser, W., and J. Kanwisher. 1961. Ecological and physiological studies on marine nematodes from a small salt marsh near Woods Hole, Massachusetts. Limnol. Oceanogr., *6*:262-270.

Wise, J. P., and J. C. LeGuen. 1969. The Japanese Atlantic long-line fishery, 1956-1963. Proc. Symp. Oceanogr. and Fisheries Resources of the Tropical Atlantic (UNESCO Publication):317-347.

Zei, M. 1969. Sardines and related species of the eastern tropical Atlantic. Proc. Symp. Oceanogr. and Fisheries Resources of the Tropical Atlantic (UNESCO Publication):101-108.

Discussion

QUESTION: What is your opinion of the value or harm in nuclear power installations along coastal waters?

DR. RILEY: I don't think that you can give a single answer to that —it depends on the situation. I would say, for example, that the thermal plants proposed for Long Island, where the water might very well be dumped into Long Island Sound, would be disastrous. There is a tendency there already for red tide production in the summer which depends in part on the heating of the surface layer, and this would intensify it. On the other hand, I could point out that there are places, such as some of our inlets on the south coast of Nova Scotia, where the water is just a little bit too cold for satisfactory growth of oysters and some other foods and where we might develop intensive fish farming with properly regulated thermal plants that would warm the water just a little in the winter and summer. Blanket generalizations are difficult.

QUESTION: Do you think that the continued large-scale use of herbicides and pesticides is a serious threat to oceanic or estuarine production?

DR. RILEY: Oh, it certainly is! The general facts of this are pretty well known. Pesticides now exist in the open ocean and in the food chain on a world-wide basis, and there are indications that this can and already does interfere somewhat with reproduction of fishes and what not.

The effect on primary production is rather less certain at this point. I don't believe that there has been any experimental work on it, at least, that I know of. Of course, my feeling on this is that it is tampering with the ecosystem and anything that we do is likely to be harmful. There are wide variations of tolerances to different forms of pesticides, of course, but even if this resulted in no more than limiting species diversity and kept up primary productivity, this still might have a very disastrous effect on other levels of the food chain.

DR. LIKENS: I wonder if you would comment from a productivity standpoint on Ryther's recent paper (Science, 166:72-76), in which he suggests that the oceans are currently being fished to approximate capacity?

DR. RILEY: I would have preferred not to comment. Everyone has made his comments on this subject, and I can remember some people saying 20 years ago that it would be impossible to as much as double the fish catch. In the intervening time we have more than doubled it, and we can certainly increase it somewhat more. However, most of the increase has been in pelagic fisheries and particularly in opening up new areas that had not been extensively fished before. We may have reached the limits of production in some areas of the North Atlantic and North Pacific. There are great stretches of ocean where there is a thinly dispersed fish fauna which would supply usable fish if we could catch them, and I can hardly agree that the open sea is quite the biological desert that Ryther makes it out to be. However, unless we develop some technological improvements that allow us to fish more effectively in areas where fish density is low, we will effect relatively little increase. It is really an engineering and economic rather than a biological problem.

QUESTION: Would you comment on the relative length of food chains in coastal and oceanic waters as described in this Ryther article?

DR. RILEY: There is a longer food chain in the open ocean if you include the whole thing, from surface to bottom. However, there is so much intermeshing in the food web, with organisms occupying different trophic levels at the same time that it becomes rather hard to estimate this or to attempt, as Ryther and some others have done, to estimate fish production in terms of primary productivity. I have seen little solid evidence to support Ryther's hypothesis that there is a distinct difference at the lower levels of the food web between coastal and oceanic waters. I would have supposed that in either situation the pelagic fish fauna would be at about the same level of the food chain and equally subject to being caught. But then in further developments of the food web, as it extends into deep ocean water, there is not enough organic matter left for effective fish production on the bottom. You could say that the food chain in shallow water is a bit longer from the standpoint of effective commercial utilization, regardless of what it may be biologically.

Measurement of Structure in Animal Communities

E. C. PIELOU*

Department of Biology
Queen's University, Kingston, Ontario

THE WORK to be described in this paper arose out of attempts to answer three seemingly unrelated questions of which only the first is explicitly concerned with ecosystem structure, one of the two topics of this volume. The questions arose during investigations of the arthropod fauna inhabiting fungus fruiting bodies; their interconnectedness became apparent gradually, and it is convenient to begin by stating them in their original, unconnected form.

Question 1: Can the notion of "the structure of an animal community" be precisely defined and quantified? In particular, can any single number—a descriptive statistic, coefficient, or index—be envisaged that would measure the amount of structure, or the degree of "structuredness," of an ecological community? It may be that great oversimplification is entailed in trying to sum up in a single number so complicated a matter as community structure. However, if one has to choose between oversimplifying field data and not simplifying them at all, the former is certainly the lesser evil. Some simplification of the data is mandatory if generality in ecological theory is sought; one can only judge whether a particular form of simplification is too extreme to be useful after it has been tested in many different contexts.

Question 2: Ecologists attempting to determine whether and to what extent different species are associated (in the statistical sense) customarily consider the species only two at a time. Greig-Smith (1964) and Cole (1957) describe methods of analysing the association among groups of three species, but dismiss as impracticable the extension of their methods to more than three species. Thus Cole (1957) writes: "If

* Present address: Department of Biology, Dalhousie University, Halifax, Nova Scotia.

one will reflect that from just four species he can select six groups of two species, four groups of three species and a group of four, the impracticability . . . becomes evident."

Nevertheless, when one has recorded whether each of a large number, say k, of species is present or absent in each of a large number of sample units, the observations should properly be summarized in a 2^k table, the k-dimensional extension of the familiar 2 x 2 contingency table. We therefore enquire whether useful, interpretable information can be derived from a 2^k table (when k is large) without first projecting it onto a space of two dimensions, or dismembering it into a number of separate 2 x 2 tables.

Question 3: Why has nobody investigated the frequency distribution of the number of species per sample unit? This distribution is immediately obtainable as soon as one has a list of the species occurring in each unit of a large sample. Few ecologists seem to have speculated on the form of observed examples of this distribution, or on what the expected distribution would be under some suitable null hypothesis. The requisite data are often collected by plant ecologists investigating species-area relations (*see,* for example, Hopkins, 1955, and Kilburn, 1966). However, when a species-area curve is to be drawn, all one wants is the *mean* number of species per sample unit for units (usually quadrats) of various sizes. The observed distribution of the number of species per unit is needed only incidentally, for computing the mean and its standard error.

The purpose of this paper is to show how the answers to these three questions are interrelated. I begin with a mathematical discussion of Questions 2 and 3 and hope to show that the results may be useful in finding an answer to Question 1.

The Interpretation of 2^k Tables

The way in which a 2 x 2 table is used to summarize the association between two species is well known. Now suppose all k species of a community are considered, and that for each species one records whether it is present or absent in each sample unit. The observations can then be tabulated (conceptually) in a 2^k table, that is, a k-dimensional hypercube containing 2^k cells each of which corresponds to one of the possible combinations of the species.

(It should be remarked here that tabulation of the data in a 2^k table is exactly equivalent to representing them by N points [one for each sampling unit] in a space of k dimensions [one axis for each species]. When, as we are here assuming, the data are qualitative, i.e., are presence-or-absence records, the "scatter diagram" yielded by the latter method consists merely of clusters of superimposed points at the vertices of a k-dimensional hypercube. The number of points at each vertex is, of course, identical with the corresponding cell frequency in a 2^k table.)

Let us take as null hypothesis that the presence or absence of every species is independent of that of all the others, and suppose we wish to test this hypothesis. It is clearly impracticable to compare the observed and expected frequencies in each of the cells of a 2^k table unless k is small (three or four) or the number of sample units exceedingly large. Except in very depauperate communities this approach is therefore unusable and we must search for some other way to proceed. No doubt several possible approaches could be devised and I have tried three. The first two of these will only be mentioned here and references made to detailed discussions elsewhere. The third possibility will then be explored more thoroughly.

The three methods are:

1. Comparison of the observed and expected frequencies in one cell of the 2^k table: Thus, one could compare the observed and expected frequencies of units containing all k species, if there are any such units; or compare the observed and expected frequencies of units containing none of the species (the "empty" units), if there are any. The method of making the latter comparison is described in Pielou and Pielou (1967). Neither comparison can be made, of course, if every sample unit contains some but not all of the species.

2. Comparison of the observed and expected number of different species combinations: To estimate the expected number (under the null hypothesis that the k species are mutually independent) it is necessary to generate, by computer simulation, a sample of "expected" 2^k tables. In each simulated table, the number of cells having a cell frequency of one or more is the same as the number of different species combinations produced by that simulation. The mean of this number, averaged over several simulations, then gives an estimate of the expected number of combinations under the null hypothesis; this estimated mean can then be compared with the number of nonzero cell frequencies in the observed 2^k table, that is, with the observed number of different species combinations. The method is described in Pielou and Pielou (1968).

3. Comparison of the observed and expected distributions of the number of species per sampling unit: At first sight this distribution seems unconnected with a 2^k table. Consideration shows, however, that it is directly obtainable from the 2^k table by adding the cell frequencies in appropriately chosen cells. Let us denote the variate, number of species per unit, by s; also, put f_i for the observed frequency with which $s = i$ for $i = 0, 1, \ldots, k$. Then f_0 is the frequency in the cell representing absence of all the species; f_1 is the sum of the frequencies in the k cells representing presence of a single species; $\ldots$;

f_i is the sum of the frequencies in the $\binom{k}{i}$ cells representing the presence of exactly i species; $\ldots$; f_k is the single cell frequency pertaining to units with all k species. The distribution of s is thus obtained by pooling the 2^k cell frequencies in a natural, nonarbitrary manner. It seems the most reasonable way of pooling frequencies, and pooling is obviously necessary when the number of separate cells in the table may be very large indeed.

It thus appears that one of the ways of coping with a 2^k table consists in answering Question 3 mentioned above.

Distribution of Species per Sampling Unit

Barton and David (1959) have shown that, given the null hypothesis of independence among the species, the distribution of s can be approximated by a binomial series. Their argument is as follows. Suppose that in a sample of N units, the number of occurrences of the ith species is $n_i = Np_i$. That is, the proportion of units containing the ith species is p_i $(i = 1, 2, \ldots, k)$. Then, examining a single unit is equivalent to performing k Bernoulli trials with, at the ith trial, probability p_i of "success" (i.e., finding the ith species) and $q_i = 1 - p_i$ of "failure" (i.e., noting the absence of the ith species).

It follows that the probability generating function of s, the number of species per unit is

$$g(t) = \prod_{i=1}^{k} (q_i + p_i t). \tag{1}$$

Now if all the species are present in the same proportion of units so that we can write $p_i = p$ for all i, this function becomes $(q + pt)^k$, the probability generating function of the positive binomial distribution. Therefore, Barton and David suggested that a binomial distribution with appropriate parameters might fit the distribution of s approximately even when the p_i were unequal.

We now require to find numerical values for the parameters of the approximating binomial. This is done by equating two different derived values of the first two moments of s. The expected distribution yields one of these derived values and the approximating distribution the other. The expected distribution has the probability generating function $g(t)$ in (1).

Thus, $E(s) = \sum_{i=1}^{k} p_i = k\bar{p}$

where $\bar{p}$ is the mean of the p_i.

Writing P and K for the parameters of the approximating binomial, which we wish to find, we have

$$k\bar{p} = KP \qquad (2)$$

Now let var (p) be the variance of the p_i.

Then $\mathrm{Var}(s) = \sum_{i=1}^{k} p_i q_i = \Sigma\,(p_i - p_i^2)$

$$= \Sigma\,(p_i - 2p_i\bar{p} + \bar{p}^2) - \Sigma\,(p_i - \bar{p})^2$$
$$= k\bar{p}\,(1 - \bar{p}) - k\,\mathrm{var}\,(p), \qquad (3)$$

is the variance of the expected distribution. Setting $\mathrm{Var}(s)$ equal to the variance of the fitted binomial distribution we now have

$$k\bar{p}(1 - \bar{p}) - k\,\mathrm{var}(p) = KP(1 - P) \qquad (4)$$

Solving (2) and (4) to yield K and P in terms of k, $\bar{p}$ and $\mathrm{var}(p)$

gives $\qquad P = \bar{p} + \dfrac{\mathrm{var}(p)}{\bar{p}} \qquad (5)$

and $\mathrm{K} = k \Big/ \left\{ 1 + \dfrac{\mathrm{var}(p)}{\bar{p}^2} \right\} \qquad (6)$

The probability generating function of the approximating binomial is then $(Q + Pt)^K$ with $Q = 1 - P$.

All the foregoing is due to Barton and David (1959) who write that the approximation "will give useful results provided $\mathrm{var}(p)$ is not too large." As far as I know no ecologists have tried fitting the distribution to their data. Of course, unless the null hypothesis is true, an observed distribution of s would not be fitted by the approximating binomial; to discover whether the approximation is satisfactory it is therefore necessary to do computer experiments and judge whether or not the distributions thus simulated are binomials with the estimated parameters. This has been tried with some field observations and yielded excellent results (see below); it was therefore possible to compare the observed and expected distributions of s. Before describing the results, we shall consider how departures from expectation are to be interpreted. What light do departures throw on species-association and, consequently, on community structure?

Associations Among Species and Community Structure

It will now be shown that the distribution of s can be made to yield an overall measure of interspecies association and it is argued that such a measure is a reasonable index of community structure.

Consider, in the first place, the change in the distribution of s that accompanies changes in pairwise associations. A 2^k table (a k-dimensional hypercube) has $\binom{k}{2} 2^{k-2}$ two-dimensional square faces each of which is a 2 x 2 table exhibiting the association between a single pair of species for some one combination of all the other species. Thus one face shows the association of species A and B in those units containing a particular combination, for instance C **D** E **F** G .... K, of the $k-2$ other species (here a bold-face letter denotes absence of the species labelled with that letter).

Suppose the observed distribution of s derived from the 2^k table is as shown in Table 1, "original frequencies." Now consider the association shown by one of the component 2 x 2 tables. Let it be the table:

		Species B		
		Present	Absent	
Species A	Present	a	b	$a+b$
	Absent	c	d	$c+d$
		$a+c$	$b+d$	

for some given combination of the other species (the lower case letters are cell frequencies).

Next let us make what may be called a *"unit increase"* in the association between species A and B (in those units containing the given combination of other species) by altering the cell frequencies of the 2 x 2 table to the following:

		Species B		
		Present	Absent	
Species A	Present	$a+1$	$b-1$	$a+b$
	Absent	$c-1$	$d+1$	$c+d$
		$a+c$	$b+d$	

Then, if m of the other k-2 species are present in the units we are considering, the distribution of s will now be as "altered frequencies" in Table 1.

Table 1. DISTRIBUTION OF THE NUMBER OF SPECIES PER SAMPLING UNIT BEFORE AND AFTER A "UNIT INCREASE" IN ASSOCIATION

Species per unit, s	$0, 1, , \ldots,$	$m,$	$m+1,$	$m+2,$	$, \ldots,$	k
Original frequency	$f_0, f_1, \ldots,$	$f_m,$	$f_{m+1},$	$f_{m+2},$	$, \ldots,$	f_k
Altered frequency	$f_0, f_1, \ldots,$	$f_m+1,$	$f_{m+1}-2,$	$f_{m+2}+1,$	$, \ldots,$	f_k

The mean of s, $\bar{s}$, is not changed since the numbers of occurrences of the k species are not changed. This also follows from the identity

$$\frac{1}{N} \sum_{i=1}^{k} if_i = \frac{1}{N} \left\{ \sum_{i=1}^{m-1} if_i + m(f_m + 1) \right.$$

$$\left. + (m+1)(f_{m+1} - 2) + (m+2)(f_{m+2} + 1) + \sum_{i=m+3}^{k} if_i \right\}$$

where the expression on the left gives the mean of the original distribution and that on the right the mean of the altered distribution.

Next consider the alteration in the variance of s resulting from the "unit increase" in association in the altered 2 x 2 table.

Denote the variances of the original and altered distributions by $var_0(s)$ and $var_1(s)$ respectively.

Then $var_0(s) = \dfrac{1}{N} \sum_{i=1}^{k} i^2 f_i - \bar{s}^2$

and $\text{var}_1(s) =$

$$\frac{1}{N} \left\{ \sum_{i=1}^{m-1} i^2 f_i + m^2 (f_m + 1) + (m+1)^2 (f_{m+1} - 2) \right.$$

$$\left. + (m+2)^2 (f_{m+2} + 1) + \sum_{i=m+3}^{k} i^2 f_i \right\} - \bar{s}^2$$

and therefore

$$\text{var}_1(s) - \text{var}_o(s) = 2/N \tag{7}$$

It follows that adjustment of any of the 2^k table's component 2 x 2 tables to give a unit increase in its association will cause an increase in the variance of s. Furthermore, every such increase will have the same numerical value, namely $2/N$.

Analogous arguments show that a similarly defined *"unit decrease"* in association in any one of the component 2 x 2 tables causes a decrease of magnitude $2/N$ in the variance of s. A "unit decrease" in association occurs when a 2 x 2 table is changed from

a	b
c	d

to

$a-1$	$b+1$
$c+1$	$d-1$

Altering the frequencies in all four cells of one of the 2 x 2 tables inevitably affects all those other 2 x 2 tables that have a pair of cells in common with it. There will be an increase in positive association (or a decrease in negative association) in one-half of the concomitantly altered tables, and the opposite change in the other half.

Now it is clear that all possible 2^k tables compatible with a given set of values of the n_i $(i = 1, \ldots, k)$ can be derived from one another by a sequence of unit changes (unit increases or unit decreases) applied to appropriately chosen component 2 x 2 tables. In particular, a 2^k table corresponding to complete independence of the species can, by successive unit changes, be converted to any observed 2^k table while at the same time the number of occurrences of each species is held constant. Each unit change is either a unit increase or a unit decrease. Let us put v for the net number of unit increases (number of unit increases minus number of unit decreases). Then, writing m_2 for the observed variance of s and $\text{Var}(s)$ for the expected variance on the hypothesis that the species are independent, it follows from (7) that

$$m_2 - \text{Var}(s) = 2v/N \quad \text{or} \quad v = N[m_2 - \text{Var}(s)]/2 \tag{8}$$

We obtain $\text{Var}(s)$ from **(3)**.

The statistic v, which may be positive or negative, is thus a measure of the net amount of positive association in the 2^k table. It depends on N, the size of the sample. A measure independent of sample size is given, of course, by v/N.

Consider now the bearing that this result has on the measurement of "structure" in an ecological community. It is surprising how often the word structure appears in the titles but not in the texts of ecological papers (to quote examples would be invidious). To me it is synonymous with "the amount of interdependence among the species." Most people will agree that a completely haphazard assemblage of species, in which none interact with any of the others is, if such a thing exists, wholly devoid of structure. Conversely, the more pronounced the interdependence among the species, and consequently the stronger the statistical association, the stronger the "structure."

This suggests that v/N may be useful as a measure of structure. Two obvious objections suggest themselves and we deal with them in turn:

1. Since v is the difference between a number of unit increases and a number of unit decreases in association, the net value could be close to zero in a community in which positive and negative associations chanced to balance each other fairly evenly. This difficulty is well known to plant ecologists studying the association between a pair of species; positive association resulting from similar habitat requirements may be masked by negative association resulting from spatial exclusion. Moreover, since a plant community usually contains only one trophic level, positive association between two species does not usually imply obligate dependence of one species on the other. By contrast, in communities of animals, especially small animals, positive associations are far more frequent than negative, for hosts and their parasites and super-parasites must be positively associated, as must carnivores and their favored prey. In communities of organisms belonging to several trophic levels, therefore, positive associations probably always dominate negative ones. "Structure," in the sense in which we are using the word here, is thus far easier to define and to measure in animal than in plant communities. Later I will remark further on structure in plant communities.

2. Even if v/N is acceptable as a measure of the association among a number of species, there is still the snag that statistical association may result from three separate causes acting singly or together. Thus there is *active* association, when species affect one another directly; *passive* association, when species are sorted into groups according to their habitat requirements in a nonhomogeneous environment; and *fortuitous* association resulting from the chance overlap of

clumps of spatially aggregated species (Pielou, 1969a, 1969b). This last phenomenon may be an important cause of association between pairs of species but becomes progressively more unlikely the more species are considered. This leaves passive and active association, which are very difficult to distinguish. A method of doing so has been described elsewhere (Pielou and Pielou, 1968) but it consumes much computer time and is inapplicable to communities occupying a large area. However, if we are willing to treat both kinds of association as properly subsumed under the description "structure," then v/N seems a useful measure.

A third seeming objection to the use of v/N as a measure of structure, namely that its value must be affected by the size of sampling unit (e.g., quadrat, water sample, soil sample) used to gather the data, is not likely to matter in practice. A statistic intended to describe one aspect of a community is rarely considered in isolation. Usually one wants to compare the structure of two or more communities, or of a single community at a succession of times, or before and after some "treatment." When this is the objective, comparable values of v/N can be obtained by using identical units for every sample. If the communities to be compared are so dissimilar that this is impossible, it is probably unreasonable to attempt a comparison.

Calculation and Interpretation of v/N

To test the usefulness of v/N as a measure of community structure, data on the arthropod fauna of the sporophores ("brackets") of a species of bracket fungus were used. Data such as these have the advantage that each individual fungus bracket is a natural sampling unit and problems connected with arbitrarily defined sampling units do not arise.

These data and the method of obtaining them have been described in detail by Pielou and Verma (1968). It suffices here to list the following six points:

1. The fungus species is *Polyporus betulinus;* it was collected from two species of birch, mostly from white birch, *Betula papyrifera,* but also some from wire birch, *B. populifolia.*

2. *P. betulinus* brackets are easily separable into two age groups: once-wintered ones which are smooth and resilient, and twice-wintered ones which are fissured and crumbling. Their faunae differ. Only data from twice-wintered brackets are presented here as their fauna is far richer in species and individuals.

3. The brackets were collected in the winter of 1967, kept in cold storage until spring, and then caged individually at room temperature.

Emerging adult arthropods (insects, mites, and spiders) were removed at one- or two-day intervals for periods of as long as six months, by which time no more emergences occurred.

    4. Although the original records list the number of individuals of each species obtained from each bracket, we shall consider here presence-and-absence records only. The numbers of individual organisms in a bracket varied enormously, from 1 to more than 1,000; these numbers no doubt depended on the sizes of the animals and of the brackets, and also on whether more than one generation was obtained from a caged bracket, as may have been true for some of the mites and the smallest insect species. These numbers are not considered further here.

    5. The brackets came from five collecting areas in mixed spruce-fir forest in New Brunswick. In what follows the collections are labelled A, B, C, D, and E. Each collecting area was in a different geographical region: areas D and E are in regions that had been sprayed with DDT annually for 14 years to control spruce budworm; areas A, B, and C are in regions that have not been sprayed. Unfortunately, comparison of community structure between sprayed and unsprayed areas has proved to be impossible as collections D and E (both from sprayed regions) come from larger collecting areas than do collections A, B, and C (all from unsprayed regions). As a result the effects of DDT-spraying and size of collecting area are confounded. Within each area all brackets found were added to that area's collection; the size of each area is shown in Table 2.

    6. To judge whether v/N is useful as a measure of structure the following comparisons have been made for data from the five areas. For each area in turn, two observed frequency distributions have been compiled. The first is the distribution of s, the number of species per bracket, taking account of *all* arthropod species. The second is the distribution of s for all species *except* the parasitic hymenoptera (these belonged to the families Braconidae, Ichneumonidae, Eulophidae, Encyrtidae, Perilampidae, Pteromalidae, Cynipidae, Scelionidae, Bethylidae, Mymaridae, and Sphecidae). Presumably the strongest positive interspecific associations in a community are those between parasites and their hosts and it was therefore expected that data from which parasites had been excluded would yield lower values of v, implying looser structure among the remaining species of the community.

    As explained earlier, v is obtained from the equation
$$v = N[m_2 - \text{Var}(s)]/2$$

Here N is the number of sample units (in this case fungus brackets) from one area; m_2 is the observed variance of s; and $\text{Var}(s)$ is the expected variance of s assuming complete independence among all the species. The value of $\mathrm{Var}(s)$ is given by (3); the equation holds whether or not the distribution of s is well approximated by a binomial distribution. However, if one wishes to judge whether the observed v differs significantly from zero, or, in other words, to test the null hypothesis that all the species are independent of one another, the observed distribution of s must be compared with the expected distribution. To find whether this expected distribution was adequately approximated by a binomial with parameters P and K as given by (5) and (6), Barton and David's approximating binomial was compared with computer-generated simulated distributions; the way in which this was done and the results obtained are described in the Appendix. It should be emphasized again that the value of $\mathrm{Var}(s)$ given by (3) is not an approximation and that therefore the value of v obtained from (8) may be accepted even if Barton and David's approximation to the theoretical distribution is not close. This theoretical distribution is needed only if we wish to fit it to the observed distribution in order to test the hypothesis that the species are all independent. In many-species communities this hypothesis seems so intrinsically unlikely as to be seldom worth testing; if a test led to its acceptance one might well suppose that a type II error had been committed. The results in the Appendix have been given to illustrate two things: the good fit given by Barton and David's distribution to simulations of the distribution of s in which independence of the species is assured; and the discrepancies between the fitted theoretical distributions and those actually observed.

Table 2 summarizes the salient results from each collection. The effect on v/N of considering each community first with, and then without, the parasitic hymenoptera is seen in the right-hand column. Exclusion of the parasites, and hence of their associations with their hosts, led to a reduction of v/N in every case.

The effect of DDT-spraying on structure cannot, unfortunately, be judged, since the effects of spraying and of collecting area are confounded. However, it is noteworthy that $\bar{s}$, the mean number of species per bracket, is greater in the two collections from sprayed regions (D and E) than in the three from unsprayed regions (A, B, and C), and this mean is, of course, unaffected by the size of the collecting area. The apparent correlation between $\bar{s}$ and v/N could be partly due to the dependence of the former on DDT spraying and of the latter on collecting area. But it would nevertheless be worth while, when additional data are available, to test the very reasonable hypothesis that abundant species occurrences and a tightly structured community go together.

Table 2. SUMMARY OF RESULTS FROM FIVE COLLECTIONS OF *Polyporus betulinus* BRACKETS.

Population	Area	Number of brackets N	Species considered	Number of species k	Number of occurrences Σm_i	Mean number of species per bracket $\bar{s} = \Sigma p_i$	$\bar{p} = \bar{s}/k$	var(p)	Var(s)	m_2	v/N
	Acres										
A	10	56	*	18	53	0.946	0.053	0.0039	0.8262	1.1578	0.17
			†	13	46	0.821	0.063	0.0050	0.7047	0.9324	0.11
B	10	83	*	43	169	2.036	0.047	0.0099	1.5157	3.1915	0.84
			†	38	164	1.976	0.052	0.0110	1.4562	3.0356	0.79
C	50	98	*	35	205	2.092	0.060	0.0142	1.4703	3.7773	1.15
			†	28	193	1.969	0.070	0.0171	1.3510	3.0297	0.84
D	100	130	*	56	380	2.923	0.052	0.0116	2.1187	4.8556	1.37
			†	44	329	2.531	0.058	0.0145	1.7469	4.0798	1.17
E	100	127	*	55	518	4.079	0.074	0.0202	2.6630	7.7733	2.56
			†	42	411	3.236	0.077	0.0224	2.0469	6.0387	2.00

* Results based on all arthropod species.
† Results obtained when parasitic hymenoptera were excluded.

Conclusions

It was argued earlier that a community can be described as strongly "structured" if there is pronounced interdependence among its member species. Such dependence will sometimes, but not always, entail the frequent co-occurrence of the interdependent species *simultaneously* and in *close proximity*. This will be so, for instance, when the organisms comprising the community are small, when they are fairly sedentary, and when they belong to several trophic levels. This description applies to the arthropod fauna of bracket fungi used as an example in this paper, and to numerous other communities of invertebrates; specialists in different groups can no doubt think of examples in their own particular fields of interest. It is in studying communities such as these that one obtains lists of the species present in each of a number of sampling units and hence, when the data are condensed, a 2^k table. Methods of analyzing 2^k tables are therefore required when one investigates the structures of communities of the kind just mentioned; that is, those from which small sampling units, each containing several species, can be taken.

Calling this kind of community Type A, we now contrast it with two other kinds, Types B and C.

Type B communities are those in which the interdependent species are not found in close proximity. An example would be any community of large, agile vertebrates, for instance an East African savannah community of predators (lions, hyenas, hunting dogs, cheetahs, etc.), prey (many species of ungulates and rodents) and scavengers (jackals, vultures, and marabous). The interdependence of the component species and thus the strength of the community's structure may be unquestionable, but it is neither feasible nor informative to sample such a community with small sampling units of fixed size. Interdependent species, notably predators and prey, are likely to be in close proximity only momentarily. Examples of this type of community on land, in freshwater, and in the sea spring readily to mind. They are often the concern of wildlife ecologists and fisheries biologists.

Type C communities are those in which the interdependent species are found in the same place but not simultaneously. Before giving examples, a digression on the structure of plant communities is desirable. Pairwise statistical associations between plant species are usually ascribed (at least tacitly) to heterogeneity of the habitat; that is, they are assumed to be "passive" (see p. 121). This assumption underlies methods of classifying vegetation that sort quadrats into homogeneous groups in which all species associations are nonsignificant or indeterminate (*see* Goodall, 1953; Williams and Lambert, 1959). If community structure is measured by the amount of statistical association among the species, and if one insists that a collection of quadrats must constitute a homogeneous group if they are to be regarded as coming from a single community, then plant communities lack structure by definition. This often seems reasonable. In most plant communities only one trophic level is present and although obligate dependence of one species on another does occur (as, for example, *Epifagus virginiana* on *Fagus* spp.) it is exceptional.

The foregoing arguments, however, take no account of vegetation mosaics which exhibit cyclical succession. Several examples of these mosaic communities were described in A. S. Watt's (1947) well known paper on pattern and process in the plant community, and a more recent account of the same phenomenon has been given by Anderson (1967). The salient feature of these mosaics is that any one point on the ground is successively covered by the different phases in repeated cycles. If such a mosaic were sampled with small quadrats, species belonging to the same phase would be positively associated and species belonging to different phases would be negatively associated but it would be absurd to assign the phases to different communities because the mosaic as a whole lacks homogeneity; taken together, the phases constitute a very highly structured community, of the kind described above as Type C.

The interdependent species are not those growing side by side at one time; they are the species that succeed one another at a particular point. Evidence for this has been given by both Watt and Anderson. Most of the mosaics described by Watt are heath communities growing on acid soils and he states that "proof of temporal sequence is found in the vertical layering of peaty remains found under the last phase." The cyclical mosaics described by Anderson are *Dryas* communities growing on raised beaches in Iceland. He found that beneath living plants of *Dryas octopetala,* in the mosaic phase dominated by this species, were the partly decayed leaves of *Betula nana* and *Vaccinium uliginosum,* species belonging to other phases which had previously grown there and whose dead leaves had remained *in situ* while the next plant in the succession occupied the ground.

Clear evidence of the association between a particular plant and a plant that earlier grew on the same spot is most likely to be discovered in regions where drought, cold, or acid soils delay decomposition of dead plant parts. But it seems likely, as Watt suggested, that cyclical mosaics may be quite a common feature of vegetation although often the evidence will not persist long enough to be recognized.

We have now considered three types of communities in which interdependence among the species, and hence community structure, are often highly developed. It is clear, however, that different methods of measuring structure need to be devised for the different types; attempts to find a single method suitable for all three are probably futile. The statistic v/N proposed in this paper is suitable for Type A communities since it is based on counts of the number of species in each of a collection of sampling units.

Recall that v/N is likely to be most useful when one wants to compare several *fairly similar communities* with one another; they might, for instance, lie along an environmental gradient, or might form a seasonal or successional time sequence at one place, or perhaps have been polluted or otherwise disturbed in varying degrees. Nothing can be said about the sampling distribution of v/N unless one first postulates some model to account for the interspecific associations in the communities being studied and usually there is no evidence in favor of any particular model. Therefore, comparisons among sample values of v/N should be planned so that distribution-free tests can be used.

It would also be interesting to search for relationships between v/N and indices measuring other statistical properties of ecological communities such as diversity and spatial pattern.

Finally it is worth noticing that the statistic v/N is derived from a 2^k table recording the frequencies of presences and absences of k species in N sampling units; or, equivalently, from a plot of N points in a k-dimensional coordinate frame. The data could equally well be represented by k points in an N-dimensional coordinate frame permitting an inverse analysis. When this is done a statistic analogous to v/N may be derived which provides a new measure of ecological diversity. Work is now in progress on this inverse approach.

Appendix

Suppose k animal species are found in N sampling units with n_i units containing the ith species $(i = 1, \ldots, k)$. That is, n_i is the number of occurrences of the ith species. We wish to obtain, by computer simulation, an empirical example of the distribution of s, the number of species per sampling unit, given that all the species are independent of one another.

Set up an $N \times k$ matrix in which all elements are zero. Each row represents one of the N sampling units (in this case, fungus brackets) and each column represents one of the k species. From the ith column of the matrix (with $i = 1, \ldots, k$) select n_i elements at random and substitute 1 for 0. Then the row totals of the matrix give the number of species per bracket in the simulated collection.

Using the data on all arthropod species (including the parasitic hymenoptera) three simulations were done for each of the five collections of fungus brackets and the first of the three is shown in Table 3 for comparison with the observed and "expected" distributions which are also shown. The expected distribution is Barton and David's (1959) approximating binomial. Its parameters, P and K, were obtained from equations (5) and (6) and are given in the table.

Since K is not an integer the approximating distribution actually consists of a mixture of two binomial series. The probability generating function of the mixture is $(1 - \delta) (Q + Pt)^{K'} + \delta (Q + Pt)^{K'-1}$ where K' is the integer part of K and $\delta = K - K'$. The variance of the mixed distribution is
$$\mu_2 = K'PQ + \delta P (1 - \delta P)$$
and therefore the difference between the variance of the expected and fitted distributions is
$$| KPQ - \mu_2 | = |\delta P^2 (1 - \delta)|$$
which has a magnitude of $P^2/4$ at most.

The bottom of the table shows, for each collection, the goodness of fit, as measured by $P(X^2)$, of Barton and David's distribution to the three simulated distributions. It is clear that in every case the fitted distribution is an acceptable approximation to the distribution that would be expected if the species were completely independent.

It is obvious on inspection that in all collections, except perhaps A, the arthropod species were not distributed independently of one another. The high proportions of empty brackets and of brackets containing many species are very pronounced in collections B, C, D, and E, and even in collection A the discrepancy between observation and expectation, though slight, is in the same direction. It is safe to conclude that v/N exceeds zero in four of the five collections and probably in the fifth also.

Literature Cited

Anderson, D. J. 1967. Studies on structure in plant communities. IV. Cyclical succession in *Dryas* communities from north-west Iceland. J. Ecol., *55*:629-635.

Barton, D. E., and F. N. David. 1959. The dispersion of a number of species. J. Royal Statist. Soc., B, *21*:190-194.

Cole, L. C. 1957. The measurement of partial interspecific association. Ecology, *38*:226-233.

Goodall, D. W. 1953. Objective methods for the classification of vegetation. I. The use of positive interspecific correlation. Austr. J. Bot., *1*:39-63.

Greig-Smith, P. 1964. *Quantitative Plant Ecology,* 2nd ed. Butterworths, London.

Hopkins, B. 1955. The species area relations of plant communities. J. Ecol., *43*:409-426.

Kilburn, P. D. 1966. Analysis of the species area relation. Ecology, *47*:831-843.

Pielou, D. P., and E. C. Pielou. 1967. The detection of different degrees of co-existence. J. Theoret. Biol., *16*:427-437.

Pielou, D. P., and E. C. Pielou. 1968. Association among species of infrequent occurrence: the insect and spider fauna of *Polyporus betulipus* (Bulliard) Fries. J. Theoret. Biol., *21*:202-216.

Pielou, D. P., and A. N. Verma. 1968. The arthropod fauna associated with the birch bracket fungus, *Polyporus betulinus* in eastern Canada. Can. Ent., *100*:1179-1199.

Pielou, E. C. 1969a. Association tests versus homogeneity tests: their use in subdividing quadrats into groups. Vegetatio, *18*:4-18.

Pielou, E. C. 1969b. *An Introduction to Mathematical Ecology.* Wiley, New York.

Watt, A. S. 1947. Pattern and process in the plant community. J. Ecol., *35*:1-22.

Williams, W. T., and J. M. Lambert. 1959. Multivariate methods in plant ecology: I. Association-analysis in plant communities. J. Ecol., *47*:83-101.

Informal Comment

In conclusion, I should like to digress a bit and comment on the contrast between *soft science* and *hard science*. Everyone is familiar with the distinction. Obvious examples of hard science are physics and astronomy, especially celestial mechanics. Thus, when astronomers talk of the return of periodic comets like Halley's, the comet always turns up at the right place, on time, and the annoying thing (to an ecologist) is that an astronomer knows his prediction will be spot on and he is so complacent about it. This is the great merit of hard science, that its predictions are *precise* and *confident*. Similar predictions can be made about eclipses, transits of Mercury, and so on. The precision and confidence also apply in "hard technology." Take the return of Apollo 13. Things went wrong but there's no denying the fact that the module splashed down where and when those in charge said it would; I don't think one could find an ecologist who could do as well.

Table 3. Observed, simulated, and expected distributions of the number of species per fungus bracket for five collections

Frequencies	Number of species per bracket (s)													
	0	1	2	3	4	5	6	7	8	9	10	11	12	13
Collection A														
Observed	24	19	7	4	2	–	–	–	–					
Simulated	19	25	8	4	–	–	–	–	–					
Expected	20.4	22.0	10.3	←——— 3.3 ———→										
Collection B														
Observed	21	17	13	16	8	3	4	1	1					
Simulated	7	21	32	15	3	3	2	–	–					
Expected	7.9	21.6	25.8	17.6	7.5	←—— 2.5 ——→								
Collection C														
Observed	19	27	19	18	5	1	6	1	1	–	1			
Simulated	9	26	26	26	7	3	1	–	–	–	–			
Expected	8.2	24.4	31.1	22.1	9.4	←—— 2.8 ——→								
Collection D														
Observed	18	17	23	28	19	15	2	2	2	2	–	2		
Simulated	3	19	24	43	25	13	2	1	–	–	–	–		
Expected	4.3	17.3	31.3	34.1	24.7	12.5	←——— 5.9 ———→							
Collection E														
Observed	9	17	15	18	21	13	8	8	5	10	1	1	–	1
Simulated	1	3	18	30	29	24	11	5	2	4	–	–	–	–
Expected	←— 6.2 →		15.2	26.1	30.3	24.9	15.0	6.6	←—— 5.9 ——→		←——— 2.7 ———→			

Table 3. (continued) Observed, simulated, and expected distribution of the number of species per fungus bracket for five collections

Collection	Parameters of theoretical distribution		Goodness of fit of expected distribution to three simulations						degrees of freedom
			(i)*		(ii)		(iii)		
	P	K	X^2	$P(X^2)$	X^2	$P(X^2)$	X^2	$P(X^2)$	
A	0.1270	7.451	1.198	0.75	4.934	0.18	3.267	0.35	3
B	0.2556	7.967	7.399	0.19	3.151	0.68	10.850	0.06	5
C	0.2971	7.041	2.837	0.72	3.570	0.61	4.118	0.53	5
D	0.2752	10.622	6.057	0.42	10.339	0.11	5.441	0.49	6
E	0.3471	11.751	7.523	0.38	2.801	0.90	6.332	0.50	7

*Values for the simulated distribution shown in the table.

There seem to be three essential requirements for successful pre-diction. You need a backlog of full, detailed, factual knowledge. You need good reasoning power. And you also need (and this is outside your control) "good behaviour" on the part of the objects whose per-formance you are predicting. Comets and planets and modules all ex-hibit good behaviour.

Now consider the contrast: look at "soft" science (we all know that it's not). The three necessary prerequisites for precise, confident prediction are difficult to obtain. First, there is a dearth of factual knowledge; or perhaps hardly a dearth, but there is so much to know that the chance of gathering in all the facts is remote. The second source of error is faulty reasoning. The third thing making precise, confident prediction so difficult is simply the vagueness and fuzziness of the things we make predictions about. This arises from the enor-mous amount of stochastic variation (chance or unpredictable varia-tion) in ecosystems—or, as engineers would call it, "white noise."

This leads to what I would call the innate unpredictability of ecosystems. One may, of course, accept defeat: If a thing is innately unpredictable, what's the use of even trying to predict it? But things are not *totally* unpredictable in ecology. Ecologists do fairly well: Those who make predictions about populations of pest species have been the most successful in predicting ecological change. But we can never escape the "white noise" problem; it is inevitable. That being so, it is particularly important that we should avoid errors that *are* avoidable. The avoidable errors are first, a dearth of factual knowl-edge; and second, faulty reasoning. We must not use white noise as an excuse for mistakes that are really due to lack of data or wrong logic.

Discussion

QUESTION: Dr. Pielou, what would you speculate would be the effect on your index of a change in sample size in a particular com-munity?

DR. PIELOU: It depends on whether you are collecting natural units (as the fungus brackets were) or are using an arbitrarily de-limited unit like a quadrat or soil sample where you can choose the size. There are bound to be effects due to sample unit size; this is a constant source of annoyance in measurement of ecological associa-tion. The only solution is to ensure that when comparisons are made, one uses units of the same size in all cases. I think a lot of trouble arises in ecology because comparisons are attempted between things that are so dissimilar that it's unreasonable to attempt a comparison.

QUESTION: If you look at a bunch of organisms just sitting out in the wild and if you call that bunch or organisms a community, you're calling it a community because they're associated. I don't see what the difference is between your measure of dependence and a measure of association.

DR. PIELOU: It's true that statistical "association" can arise either because two species merely happen to share the same habitat, or because they actively interact. It is often very difficult to distinguish the true cause of association. A mathematical test applied to a 2 x 2 table tells nothing about causes. My husband and I in a joint paper (J. Theoret. Biol. [1967] *16*:427-437) did describe a method by which one can sometimes judge whether an observed association is due to true interdependence or merely to the co-occurrence of species with similar environmental requirements. However, in today's discussion, I have lumped the two kinds of association together because the task of distinguishing between them is very laborious even when it's possible.

QUESTION: Do you think the difficulty of distinguishing the effects of these two causes would affect your association index?

DR. PIELOU: No. It would introduce more white noise but whether this would make the measure of structure valueless I can't say, because it's still largely speculative.

The whole purpose of this approach was to devise a *single* measure of the association among k species, in the same way as mean square contingency, for example, can be used for only two species. If one wants to consider the mutual dependence of only two or three species one can use partial correlations as described in Cole's 1957 paper. That would be the way to proceed if one were interested in only a very small subset of the total number of species. However, I wanted to take account of all species, no matter how many, and obtain a single measure that was deliberately designed to confound many different interactions.

QUESTION: Have you thought of the range of possible values of v/N? What would be the range of possible values and what's going on when it gets to be very large?

DR. PIELOU: The answer to your first question is no, and from that all else follows. Sorry to be of no help.

QUESTION: I wonder if you'd be willing to discuss the speculation that interdependence, or "structure" leads to ecosystem fragility while diversity yields stability?

DR. PIELOU: I'm not convinced that there is a one-to-one relationship between stability (however you would like to define that) and diversity. All the arguments in the literature on the topic have been based on particular examples, and for any example that a person in one camp produces showing a positive correlation between stability and diversity, someone in the opposite camp can easily find a counter-example. So it goes—it's like watching a tennis match: I don't know of any logically incontrovertable argument that would lead you to associate stability and diversity. One might say that "on balance" and "on common sense grounds" one would expect this or that, but it isn't susceptible to logic and doesn't really lead anywhere.

QUESTION: Could I ask then what is the relationship between your index of structure and an index of diversity?

DR. PIELOU: There's no connection between this index of structure and any existing diversity index. The index of structure was arrived at by plotting the n points (representing samples) in a k-dimensional coordinate frame in which each axis represents a species. Now there's nothing to stop you, if you have a mind to, from reversing the whole process and portraying your results as k points (to represent species) in an n-dimensional coordinate frame. Each axis of the frame then represents a sample unit. If one now derives an entirely analogous "index" it may turn out to be interpretable as a diversity index.

Evolution of Natural Communities*

R. H. WHITTAKER
Section of Ecology and Systematics
Cornell University, Ithaca, New York 14850
and
G. M. WOODWELL
Biology Department
Brookhaven National Laboratory, Upton, New York 11973

THE STUDY OF the evolution of the natural communities that occupy the earth's surface might well be a major topic in science. It is, instead, the subject of a remarkably small literature. In making this statement, we are limiting our concerns in some respects, omitting description of past communities and their environments (Ladd, 1957; Imbrie and Newell, 1964), and omitting as well the evolution of the biosphere (Berkner and Marshall, 1964; Laporte, 1968; Cloud, 1968; Simpson, 1969). Our emphasis is on the evolutionary continuity of individual natural communities and evolution of the characteristics of those communities.

Let us suggest a few principles of community evolution that should be important:

1. All species evolve in communities in interaction with other species. The community is the context of species evolution.

2. The evolution of a community must entail "parallel" or co-adaptive evolution of the community's species. The community is an assemblage of interacting and coevolving species.

3. Hence communities must change and may diverge in structure and function from ancestral communities through evolutionary time. Communities, like species, should be related to one another by phylogenies.

---

* This paper was delivered at the Colloquium by Dr. Whittaker.

4. Through this evolution there will appear adaptation to environment for the community as well as for the species. From this evolution results a mosaic of communities adapted to the mosaic of the world's environments.

5. Community level characteristics also will evolve. One need not conceive of the community as a superorganism (Clements, 1916, 1936; Phillips, 1934-5) to think that these emergent characteristics should correspond to some of those of organisms (cf., Allee *et al.*, 1949: 440). They should include individual growth and maturity, structural differentiation, energy flow and materials turnover, steady state function and homeostasis, adaptive optimization, and organization.

With these five principles we may appear to have covered the topic and might conclude our discussion. The principles have, however, more of the quality of bromides than of penetrating interpretations. We propose now to examine these principles under several topics that are widely accepted as community attributes: diversity, pattern, physiognomy, succession and climax, productivity and biomass, and nutrient cycling.

Species Diversity

We begin with a quantum increase in the community—the addition to it of one species by the extension of that species' range. The habitat of the community is already occupied by other species. Let us accept the principle of Gause or of competitive exclusion (Gause, 1934; Hardin, 1960; Slobodkin, 1962). The new species may displace a competitor already present, may survive because it differs in niche from other species present, or may evolve difference of niche from the species most closely competitive with it. The divergence of niche can occur by differential selection among individuals whose genetic characteristics make them more closely, or less closely competitive with another species in the community. Larger proportions of the latter individuals survive and reproduce to transmit their characteristics to future generations (Wallace and Srb, 1961; Whittaker, 1970a). The new species in any case fits into the community by difference in position, resource use, or interaction with other species in the community—by difference in niche. The community is a record of many successful species additions accumulated through evolutionary time, edited by extinctions. The process makes the community a system of interacting, niche-differentiated species.

An extensive literature exists on theory and mathematical description of relations between species within the community. We shall, however, limit ourselves to the over-all result of the additions of species. This is "alpha species diversity," the richness in species of a particular sample or community (Whittaker, 1960, 1969). It is one aspect of community-level structure; some points regarding this structure are:

1. Species diversity increases, by addition of species and elaboration of niches, through evolutionary time.

2. The extent of the increase in diversity is affected by evolutionary time and by stability and favorableness of environment (Dobzhansky, 1950; Fischer, 1960; Whittaker, 1965, 1969; Sanders, 1968, 1969). Diversity is affected by evolutionary time through which species additions to the community have occurred. Stability and favorableness of environment affect the number of species that have been able, during that time, to adapt to and survive in a given environment. They affect also the degree to which selection may direct species evolution toward niche differentiation and survival in relation to other species, permitting increase in diversity, rather than toward adaptation for surviving environmental hazard.

3. Increase in species diversity is potentially a self-augmenting process, without a limit or ceiling (Hutchinson, 1959; Whittaker, 1969). This seems to be the case for vascular plants and for insects (Whittaker, 1970b). The niches of birds, however, relate to the structure of the vegetation and to broad categories of food. The number of distinct niches that can be defined on the basis of these is limited, and there is consequently a limit to the number of species of birds that a community of a given structure can support (MacArthur, 1965).

4. The relative importance of species in communities—as a measure of the way species have divided resources and fractions of community productivity among themselves—may be treated as dominance-diversity curves (Whittaker, 1965, 1969, 1970a). These curves are of interest for the range of different forms they take, presumably expressing (though in a somewhat ambiguous way) different types of competition and division of resources among species.

5. We note, finally, that this aspect of structural differentiation of the community is a product of the way species evolve in relation to one another. It does not appear that diversity or importance-value relations are affected by differential selection applied to communities as wholes.

Community Pattern

The new species added to a community must find its habitat as well as its niche. Habitats are in many cases continuously related along environmental gradients. The adaptive problem for the species then entails finding both a niche within communities and a distributional relation to the environmental gradients along which the communities occur. These two aspects of its adaptation are closely related. Species that are closely similar in their niches can coexist in an area by occupying different positions along environmental gradients. If the new species encounters an old species of similar niche and habitat requirements, the two populations will tend to diverge in their habitats, as well as their niches. Competition may be reduced by either niche differentiation or habitat differentiation or—and perhaps this most commonly—by adaptive shifts in both habitat and niche at the same time (Whittaker, 1969). Reduction of competition by difference in habitat implies that species evolve toward scattering of their population centers along environmental gradients. The fact that the species also differ in niche permits their populations to have broadly overlapping distributions along the gradient.

Species do not evolve toward formation of coherent groups, bound together by coadaptations that result in similarity of distribution. They evolve instead toward diversity of distributions. Along environmental gradients they form the kind of population continua that are observed in gradient analysis, with broadly overlapping, bell-shaped population curves with scattered centers (Whittaker, 1951, 1956, 1969, 1970a). This is a result that seems quite at variance with an uncritical expectation of how species and communities might evolve. It is consistent, however, with the principle of species individuality (Ramensky, 1924; Gleason, 1926), the assertion that each species is distributed according to its own genetics, physiology, population dynamics, and way of relating to all factors of environment including other species. Because of this individuality of adaptation, species are "individualistic" in distribution—no two alike (apart from some obligate symbionts).

Through evolutionary time additional species can fit themselves in along an environmental gradient by varied combinations of niche and habitat difference from species already there. Addition of niche-differentiated species to particular communities increases their alpha diversities or richness in species. A different aspect of diversity is affected by adding habitat-differentiated species along an environmental gradient. Here competition narrows amplitudes of populations along the gradient and evolution increases "beta diversity"—the degree of change in community composition along environmental gradients (Whittaker, 1960, 1967). Thus MacArthur (1965) has observed that whereas alpha diversities of bird communities appear to be similar in temperate and tropical communities of similar structure, beta diversities of bird communities increase from the Temperate Zone into the Tropics.

This aspect of evolution makes it important to consider two further points that affect our introductory principles. First, we must consider the thesis that natural communities form complex and predominantly continuous population patterns in relation to patterns of environmental gradients (Whittaker, 1951, 1956, 1967). The thesis results from research in gradient analysis that studies relations of populations and community characteristics to gradients of environment. An "ordination" represents in the abstract some aspects of such a pattern (Whittaker, 1967). The traditional conception of a mosaic of communities corresponding to a mosaic of habitats is not simply false (especially for man-made mosaics), but is a less realistic, less understanding conception.

The second implication for our principles involves the evolutionary "individuality" of species. Through evolutionary time, species can change their adaptations and associations with one another (Mason, 1947; Whittaker, 1957). Although species adapted to the same kinds of environments may tend to stay together in evolutionary time, the longer that time the more likely they are to separate by different adaptive responses to changing environments and to occur with new associates. A community is not simply the lineal descendent of some past community; it comprises species of diverse evolutionary histories in diverse past communities. It is possible to construct histories of broad patterns of differentiation of communities in relation to changing climates, as Chaney (1940, 1947), Axelrod (1958), Braun (1947, 1950), and others have done. But as involves species composition, there is not a true divaricate phylogeny of communities. The difficulty is not the kind of distinct reticulateness now accepted for many plant genera—in which one plant species has evolved from two or three other species. Communities are related by a blurred reticulateness of many intersecting strands (i.e., species) relating a present community to many past communities. The evolutionary relationship is more like the genetic relationship of human families to one another than of species to one another. This may be an important reason, along with the limitations of fossil records, that detailed investigations of community phylogeny in terms of continuity of species composition are so small a part of the great endeavors of paleontology.

Physiognomy

Dominance-diversity curves are one expression of the structure of the community, the division of resources among its species; physiognomy, a more traditional concern of plant ecology, is another.

A plant community is a mixture of growth-forms; community structure is determined by the character and proportions of those growth-forms. Differences in growth-form among the plants of a community are incomplete expressions of differences of niche—differences in resource use, manner of competition, and seasonal timing. The new species added to the community must fit itself in by some distinctiveness of niche, and this distinctiveness may represent addition to the community of a growth-form in addition to the range of growth-forms already present. Structural differentiation then, like species diversity, may increase through evolutionary time.

Physiognomy is an evolutionary product expressing, as does an organism's structure, adaptation to environment. Environment sets constraints on growth-forms of species and on the physiognomy of the community. Some growth-forms are excluded: tree ferns, mangroves, and stem succulents do not occur on the Oregon coasts. Some growth-forms are especially adapted to take advantage of the resources of a given environment, and species with these growth-forms become dominants of communities. Other growth-forms include species that not only tolerate the environment, but exploit the new conditions set by the dominants. Such species accumulate through evolutionary time to make up the community's spectrum of subordinate growth-forms.

From these responses of growth-forms to environments results the physiognomic convergence of communities. Communities of similar physiognomy appear in the similar climates of different continents; a grouping of communities of such convergent structure is a formation-type or biome-type (Dansereau, 1957; Odum, 1959; Whittaker, 1970a). Thus communities of a most distinctive type, the giant temperate rainforests, appear in highly humid, temperate climates of the northwestern United States and eastern Australia. The convergence of community structure on different continents is imperfect because of the different evolutionary stocks available. In Australia, eucalypts form a broadleaf rainforest in contrast with the needleleaf rainforest of the United States; and drought-adapted eucalypts apparently form woodlands in Australia in climates that on other continents would support grasslands.

The convergence is nonetheless an important part of the biogeographic design of the world's vegetation. Much of the structural response of vegetation to climate can be treated in terms of formation series (Beard, 1944, 1955; Whittaker, 1970a)—sequences of formations, or structural types of communities, occurring along major environmental gradients. Plant growth-forms do not in general exclude one another along sharp boundaries, but overlap in distribution. Community physiognomy for the most part changes continuously along environmental gradients. A formation series is usually a physiognomic continuum, fractions of which we choose to recognize as formations (Beard, 1955; Whittaker and Niering, 1965, 1968). (There are, of course, some more abrupt physiognomic boundaries, among which the fire-sharpened, prairie-forest edge in the United States is notable.) It is convenient to discuss the vegetation of the world in terms of "formation-types" or "biome-types," but it may be conceived also as a pattern. As the species of a given landscape are variously combined into communities to form a population pattern in relation to the environmental pattern, so growth-forms, each distributed in its own way, are variously combined into communities to form the physiognomic patterns of vegetation in relation to climatic patterns of the continents.

Much of what Chaney, Axelrod, Braun, and others describe as vegetational evolution is not a phylogeny. It is progressive physiognomic differentiation in relation to climatic differentiation. The transcontinental Arcto-Tertiary forests or the Madro-Tertiary vegetation of the Southwest may always have shown significant geographic differentiation in species composition (Whittaker, 1957, 1961a; cf. Wang, 1961). In the details of species adaptations and associations, the evolution from these Tertiary patterns to present vegetation was complex, multidirectional, reticulate, and obscure. Superimposed on this complexity is a more easily conceived, more effectively divaricate physiognomic divergence—a differentiation in response to intensified climatic gradients, that intensified the physiognomic differences within Arcto-Tertiary and Madro-Tertiary vegetation patterns.

In discussing physiognomic evolution we are not concerned with evolution through selection of communities as communities. Selection acts instead on genetic characteristics and survival of species populations. Dominants and subordinates of different growth-forms survive, accommodate to one another, and share resources in communities. Physiognomy, like species composition, is an integrating product of this species selection and coadaptation.

Succession and Climax

Physiognomic gradients appear also in successions. A succession after fire on Long Island (Whittaker and Woodwell, 1968, 1969) may proceed through stages: annual herbs, perennial herbs including *Carex pensylvanica*, vacciniaceous low shrubs, high shrubs (*Quercus ilicifolia*) mixed with young pines and oak sprouts, young forest of pine (*Pinus rigida*) and oak (*Quercus alba, Q. coccinea*), and mature oak forest. One can think of the succession as a formation series along a time gradient. The manner in which species populations rise and fall and replace one another along this time gradient appears to be similar to the manner in which species populations change along a habitat gradient. Population change through the course of succession appears in many cases, though not all, to be continuous. Population gradients in succession may differ from those along habitat gradients, however, in greater place-to-place irregularity. The timings of particular species in succession, and the species composition of successional stages, seem less predictable from one succession to another (in similar habitats) than are species positions from one habitat gradient to another (Whittaker, 1953). Some species are adapted to successional communities and are absent from, or occur at lower densities in, stable or climax communities. Successional time is thus a third major direction (with niche and position along habitat gradients) in which species may differ in ways that reduce competition between them.

The succession may also be a physiognomic continuum of changing dominant growth-forms and combinations of growth-forms. Certain developmental trends characterize many successions, although exceptions to almost all these trends can be found (Whittaker, 1953; Margalef, 1963; Odum, 1962, 1969). Through succession on land there tends to occur: (a) progressive increase in height of the dominant growth-forms, and hence of the communities; (b) increasing diversity of growth-forms and stratal differentiation of communities; (c) increasing species diversity; (d) progressive soil development with increasing depth, organic accumulation, and horizon differentiation; (e) increasing community productivity and respiration, and progress toward balance of these; (f) increasing biomass, and increase in the stock of nutrients held in the biomass; (g) increasing modification by the community of its internal environment; and (h) increasing relative stability of species populations. Some of these reverse themselves in later stages of many successions. The climax is not to be defined by maximum productivity and species diversity, but will usually be characterized by maximum biomass accumulation and by steady states of species populations, productivity (approximate balance of gross photosynthesis and total respiration), and nutrient circulation.

It is possible by long-term disturbance (overgrazing, introduction of toxic materials, ionizing irradiation) to produce reversal of most of these trends (Woodwell, 1970). Such retrogression is illustrated by the gamma irradiation of the Brookhaven forest on Long Island (Woodwell, 1962, 1967; Woodwell and Whittaker, 1968a). With increasing intensity of radiation exposure the communities retraced (approximately) the physiognomic gradient already described: from forest to shrub stage, to *Carex* meadow, to mosses and lichens, and to a community limited to crustose lichens and microorganisms. Accompanying these changes were reductions in plant height, stratal differentiation, species diversity, productivity, and living biomass. The changes include also a drastic modification of the internal environment of the community. Two observations from this retrogression are of particular evolutionary interest.

First, there are the parallel characteristics of ecoclines (gradients of communities and their environments) or formation series in diverse circumstances. Successional series in some respects resemble formation series from more adverse to more favorable environments. The Brookhaven retrogression gradient on the one hand resembles a succession in reverse, and on the other resembles in major plant types a formation series from the eastern forests to the high Arctic (Woodwell, 1967).

Second, characteristics of species relate to these diverse circumstances. There is some tendency for the same species and plant forms to be adapted to environmental rigor, successional roles, and the biologically unusual stress of gamma irradiation. Tolerance of irradiation (and, apparently, other stresses) in general decreases with increasing size of plants and with increasing mean chromosome volume (Woodwell, 1967).

Organisms and communities both have developmental relations that can be described as "growth" and "maturity." The difference, beyond these descriptive words, is profound. Complex, highly evolved, encoded instructions are inherited by the organism and govern the direction of its development and the characteristics of its maturity. Successions, in contrast, give the impression of a hit-or-miss interplay of species populations that reach and can tolerate the habitat, replacing one another in a way that may be only loosely predictable, until the successional process runs down into a self-maintaining combination of species, a climax. There is no inherited, determinate end point; characteristics of the climax are determined jointly by the resources and constraints of environment and the characteristics of particular species as they compete in that environment. To reach the climax the succession may be either "long" (involving a number of dominant populations and distinguishable stages), or "short" (in which the climax species enter the bare area in the first stage, and mature to form the climax, as occurs in desert and tundra) (Muller, 1940, 1952; Shreve, 1942). Succession and climax are important expressions of evolution and adaptation of species and communities. But the community level characteristics of successions and climaxes are consequences of the evolution of species and their interaction and function in relation to environmental constraints and resources.

Productivity and Biomass

It is clear that differences in productivity underlie many differences in physiognomy. The change in stature of the community from forest through woodland and shrubland or grassland to desert is in part an expression of environmental constraints on community structure exerted through their effects on productivity. The difference in physiognomy between a deciduous broadleaf forest and an evergreen-sclerophyll forest in temperate climates of the same annual rainfall reflects differences in seasonal cycles of productivity. The depression of structure along the gradient of increasing irradiation at Brookhaven involved reduction of community productivity along with the reduction of the stature of the plants that the productivity supported (Woodwell, 1967).

More directly, differences in physiognomy involve differences in community mass. Margalef (1963) has emphasized the ratio of production to biomass and its decrease during succession. The inverse ratio of biomass to production expresses more directly what is of ecological interest. For this expression we may use the "biomass accumulation ratio"—the total biomass or dry weight of living and dead tissues of organisms, over the net annual productivity in dry weight, for the same community area (Whittaker, 1961b, 1966). This ratio is of interest in relation to succession and climax, stability and time binding, structure, diversity, and nutrient function.

The forest accumulates biomass, woody tissue to support its photosynthetic apparatus, during succession. Biomass accumulation ratios increase from close to 1.0 in an annual herb stage to probably 2-5 in perennial herb and low shrub stages, to 5-10 in high shrub and young forest stages, to 30-50 in many mature forests (Whittaker, 1966). It is not productivity but biomass—or its accumulation in relation to productivity—that reaches a maximum in the climax of most terrestrial successions. Plankton communities in contrast are unstable and changeable suspensions subject to rapid turnover of plant cells by grazing and sinking; they have no such possibility of biomass accumulation. Plankton biomass accumulation ratios appear to be inverse to values for forests and of the order of 1/20 to 1/50. One can describe successions involving increasing biomass accumulation and species diversity for experimental plankton communities (Margalef, 1963), but it is difficult to apply the climax concept as discussed here to aquatic natural communities.

The forest community has a stability in consequence of its long-lived dominant populations that is in contrast with the rapid fluctuation of species in plankton communities. The forest community has a "time binding" character, embodying in its structure and composition the effects of past years' environments on populations and their inter-actions. The forest has a stable, complex structure of different plant tissues that offers a number of niches to consumers far exceeding the number of plant species. Pielou's example of many arthropod species in bracket fungi (this volume) is a dramatic illustration of the rich-ness in species that specialization for use of small parts of the complex structures of forests has made possible for insects. It is likely that the ratio of consumer species numbers to producer species numbers is much higher in the forest than in the plankton (Gallopin, pers. comm.).

Fire has marked effects on biomass accumulation and structure. Between forests' too wet to burn and deserts too open in structure to carry fire, various kinds of transitional communities occur—wood-lands, shrublands, and grasslands. These are alternative physi-ognomies for communities in climates of similar rainfall and tem-perature. They are alternative evolutionary responses to the effects of fire as influenced by climate, soil characteristics, topography, and fre-quency of burning. Grasslands in particular are evolutionary responses to fire, with biomass accumulation and structure lower than in for-ests and desert shrublands of wetter and drier climates. Grasslands have evolved community designs, with structure adapted to fire and with means of protection against grazing (including the deposit of abrasive silica particles in their leaves) quite different from those of woody communities (in which other secondary substances may provide protection against grazing) (Whittaker, 1970b). Evolution of grass-lands offered an evolutionary challenge that was met by evolution of animal communities with responses to vegetation structure and use of plants for food quite different from those of forest animals (Allee, et al. 1949; Odum, 1959).

We may mention two further implications of structure for pro-ductivity. Terrestrial communities, especially forests, have foliage of different species staged in depth, intercepting light of different intensi-ties to which the various species may be adapted. Our new species must fit itself in, in relation to this light intensity gradient and inor-ganic nutrient needs. Theoretically, a larger number of species differ-ently adapted to light and nutrient use could have a higher productivity (for a stable community) than a smaller number of species less effec-tively complementing one another in resource use. There is only lim-ited evidence from experiments with cultivated species (Harper, 1967, 1968) that plants complement one another in such a way as to produce more in mixed stands than in single species stands. Apart from the contribution of nitrogen-fixing plants to production by other plants, there is apparently no effective evidence that complementation increases productivity in natural communities. Salt marshes, among the most productive of communities, are often virtual monocultures. The community-level benefit of species diversity for productivity would be a fine evolutionary principle—but it is not necessarily true.

Nutrients and Chemical Function

The other implication of structure is for nutrient function. It is through effects on nutrient circulation that contrast of structure implies contrast of production in forest and plankton. The plankton, with its minimal structure, does not effectively retain inorganic nutrients. The continual settling of plankton cells and particles carries the nutrients in these downward out of the lighted zone of the water. The nutrient pool of the plankton in the lighted zone is consequently reduced to a low level, at which loss of nutrients by sinking is balanced by upward transfer of nutrients by water turbulence (Steele, 1958; Riley, 1965) and migration of organisms. To this low, steady-state level of nutrient supply corresponds a low level of productivity (Ryther, 1963; Strickland, 1965).

The forest, in contrast, develops a structure holding a large pool of nutrients in its tissues and dead organic matter. Gradual release of nutrients from litter, combined with relatively rapid uptake of released nutrients by roots and fungi, holds the greater part of the nutrient stock against loss. In certain tropical forests nutrients appear to be held in tight circulation in the communities occupying the surfaces of strongly leached soils (Went and Stark, 1968). The significance of forest nutrient retention has been dramatically shown also in the contrast of cut and uncut watersheds in the Hubbard Brook study described by Bormann et al. (1968) and Likens et al. (1970 and this volume). The accumulation of biomass and nutrients during succession permits the development of a massive woody forest structure supporting the photosynthetic apparatus of leaves arranged in depth and an annual productivity approaching one order of magnitude higher than that of the open ocean plankton (Whittaker, 1970a). The contrast is an evolutionary response to a stable surface vs. unstable suspension in water.

It is reasonable to look for differentiation in nutrient function at the level of the species. Phytoplankton species may differ in: (a) relative requirements for different inorganic nutrients; (b) relative tolerance of one another's inhibitory ectocrines; and (c) relative need for, or use of, organic nutrients. Plankton cells are leaky systems that release significant fractions of their organic productivity into the water, and many phytoplankton take up and use (and some require) organic compounds from the water in supplementation of their own photosynthesis (Lucas, 1947; Hartman, 1960; Fogg, et al. 1965; Danforth,

1962; Droop, 1962). The plankton species are thus part of a network of biochemical exchanges. This three-fold matrix of chemical differentiation and interrelation may contribute much, along with other effects discussed by Hutchinson (1961), to the paradox of the plankton—the puzzling richness of the plankton in photosynthetic species apparently competing for the same light and nutrients.

We may expect chemical differentiation of species in the forest also. Morphological evidences of differences in nutrient function appear among the major species in the Brookhaven forest (Whittaker and Woodwell, 1968, 1969). Vascular plants are, we believe, chemically protected by their secondary substances against consumption and infection (Fraenkel, 1959; Whittaker, 1970b). Animal and saprobe species dependent on higher plants have evolved into chemical accommodation with them. A forest may be thought of as a set of loosely bounded biochemical species complexes, each comprising a plant species and the animal and saprobe species directly or indirectly related to it. Such complexes imply occupation by insects of parallel niches— niches that are alike, say, in the plant part consumed and manner of consumption, but differ in the chemistry of plant species. Chemical differentiation may thus have much to do with diversity of the land plants and insects (Ehrlich and Raven, 1964; Whittaker, 1969, 1970b; Whittaker and Feeny, 1971).

Thus we come back to the new species entering a community with which we started. That species will fit itself in by adaptive differences from other species—difference in position along environmental and successional gradients, in spatial and temporal position in the community and response to environmental variables, in manner of using light and nutrients for production, in biochemical characteristics and accommodations, and in means of regulation by resources and effects of other species. The trend of evolution is toward such differentiation in species characteristics as a given environment permits, and thereby toward complexity of structure (as based on diversity of species and growth-forms) of the community.

Emergence

Community complexity and diversity are thus products of evolution. We should like, however, to reconsider the matter of emergent community characteristics.

Species diversity, species composition, importance-value relations, and physiognomy are such community-level characteristics. But these also are community-level products of evolution and interaction at the species level. They are consequences of the way species adapt to environment, differentiate in niche and habitat, and divide among them the resources of the community. We know of no selective mechanism, acting at the community level on communities as wholes, toward higher diversity, or toward physiognomic characteristics that are not determined by the selection of the dominant and other major species.

There is the question of the maximization of production, or the optimization of community characteristics related to it. Studies of temperate forests in the Smokies and at Brookhaven suggest convergent characteristics—net production of 1,200-1,500 dry $g/m^2/yr$, leaf production around 400 $g/m^2/yr$, surfaces of leaves (deciduous, one side) of 4-6 m^2/m^2, bark surface of 2.0-2.5 m^2/m^2, mature biomass of 30-60 kg/m^2, etc. (Whittaker, 1966; Whittaker and Woodwell, 1967, 1969; Woodwell and Whittaker, 1968b). It can be shown that these are not maxima but norms that are exceeded in favorable circumstances (Whittaker, 1966). For most forests they may be optimum dimensions in a limited sense. There may be diminishing photosynthetic advantage from the production of leaves above about 400 $g/m^2/yr$, and biomass and bark surface above 60 kg/m^2 and 2.5 m^2/m^2 may shift the balance of respiration to photosynthesis toward the unfavorable for forests that are not exceptionally productive. These dimensions are not, however, inherited limits implying determinate growth of the community as such; they are responses of plant populations to environmental limits. In a forest strongly dominated by one species, the community dimensions are in large part products of that species' population density and its evolved pattern of average dimensions reached by mature individuals. (The average dimensions involve interactions of size and form with density and variation of these interactions with environment.) In a mixed forest community dimensions are the sums of dimensions achieved by a number of species that share, and by competition limit one another's portions of, the resources of environment.

There is in the community no center of control and organization such as there is in the organism, and no evolution toward a central control system. The organization of the community—its pattern of self-maintaining functional differentiation—is maintained by the genetics of species, by way of the competition and other population interactions of these species. Community organization is a result of species evolution and species behavior. This organization, and the related spatial organization of communities as complex population patterns, are distinctive (Whittaker, 1969). Community organization, with its quality of "loosely ordered complexity" (Whittaker, 1957), has no really good analogy among other living systems.

There are, finally, the important ideas of stability and homeostasis. The stability of the community is quite different from the homeostasis of the organism and involves four processes, at least:

1. SPECIES AS COMPENSATION DEVICES. If chestnut (*Castanea dentata*) disappears from an eastern forest, oaks and other species replace it in the canopy. Forest productivity, biomass, and structure return to their norms; forty years later one might not recognize that chestnut had been there (Woods and Shanks, 1959; Slobodkin, *et al.* 1967). Such compensations by species replacements might be thought—though the analogy is not close—a kind of buffering; they are changes in the relations of particular components by which change in the characteristics of the system is reduced or prevented.

2. DENSITY-DEPENDENT POPULATION REGULATION. Some of the factors affecting populations act with increasing effectiveness to increase deaths or decrease births as population density increases. These density-dependent factors act to limit population growth and to damp population fluctuations; they can produce relative stability of populations from year to year. Despite the fact that the stability of populations in the field is so clearly imperfect (Ehrlich and Birch, 1967), most natural species populations appear to be relatively stable. Presumably a variety of density-dependent factors including effects of resource limitation and interactions between species (parasitism, predation, competition, etc.) are responsible for stability (Solomon, 1949; Odum, 1959; Hairston, *et al.* 1960; Slobodkin, *et al.* 1967). Because a larger number of species in the community implies a larger variety of species interactions of which some are density-dependent, stability of communities should tend to increase with increasing species diversity (MacArthur, 1955; Margalef, 1969). The relation is not a strict and necessary one, but there should be a tendency toward evolutionary (as well as successional) increase in community stability.

3. ENVIRONMENTAL CONTROL BY THE COMMUNITY. The Hubbard Brook study has shown the effect of the community in governing its nutrient pool and stabilizing the composition of the groundwater leaving it (Bormann, *et al.* 1969; Likens, *et al.* 1970). The community tends to damp environmental fluctuation, to stabilize its microclimate and environmental chemistry. Such effects are roughly proportional to the community's biomass and coverage.

4. STEADY-STATE FUNCTION. Chemical stabilization implies steady states, development of the community such that output, as a rate value applied to the community pool, balances input. The steady state applies also, for stable communities, to energy flow and to turnover of individuals through populations. This tendency toward the development of

steady-state functions, applying to different processes on different levels, is one of the most general characteristics of ecosystems. The most familiar model is one in which, for a given input rate I, pool magnitude increases until output (a constant rate fraction k, times pool magnitude M) equals input, $I = Mk$. (But both I and k may be variables in successions.) Thus forest production, biomass, and decomposition are interrelated, for example 1,500 $g/m^2/yr$ = 60 kg/m^2 x 0.025/yr, for a biomass accumulation ratio of 40. The steady-state magnitudes of communities seem not to be directly selected for, but to be resultants of processes which tend, by some mode of growth until output and input are equal, to balance themselves.

We conclude that community characteristics are to be understood as cumulative effects of species evolution and multiple steady-state processes. We should like, though, in scrutinizing some traditional ideas to comment on one—the balance of nature. This expression is as common in the popular literature of conservation as it is unpopular among scientists—though it may be conceived as the tendency of populations and communities to compensate for perturbations (Slobodkin, et al. 1967).

What is its truth? We suggest that it is the intuitive perception of the steady-state function of populations, ecosystems, and the biosphere, by naturalists to whom the steady-state concept as such was mostly unknown. It is the inference that, underlying the relative constancy of the natural world from year to year, there must exist functional balances and regulatory processes. The perception—of communities as tending toward self-stabilization involving steady-state function, and of the biosphere as a great steady-state system of interconnected steady-state systems—was not false. The emergence of loosely ordered and imperfect steady-state functions is a most general feature of the evolution of ecosystems. A technological civilization might make no greater mistake than to think that the balance of nature is irrelevant to human affairs. As Golley (in this volume) has observed, it is in our interest to protect what is left of this balance if we can, for we cannot adequately foresee the disadvantages to us of further disturbances. Man stands now at a turning point in history where he has the power to change not only the functions of particular ecosystems, but the characteristics of the biosphere as a whole. It seems a fateful turning.

Acknowledgment

Research carried out at Brookhaven National Laboratory under the auspices of the U. S. Atomic Energy Commission.

Literature Cited

Allee, W. C., A. E. Emerson, O. Park, T. Park, and K. P. Schmidt. 1949. *Principles of Animal Ecology.* Saunders, Philadelphia.

Axelrod, D. I. 1958. Evolution of the Madro-Tertiary Geoflora. Bot. Rev., *24:* 433-509.

Beard, J. S. 1944. Climax vegetation in tropical America. Ecology, *25:*127-158.

Beard, J. S. 1955. The classification of tropical American vegetation types. Ecology, *36:*89-100.

Berkner, L. V., and L. C. Marshall. 1964. The history of growth of oxygen in the earth's atmosphere. *In* P. J. Brancazio and A. G. W. Cameron [eds.], *The Origin and Evolution of Atmospheres and Oceans,* pp. 102-126. Wiley, New York.

Bormann, F. H., G. E. Likens, and J. S. Eaton. 1969. Biotic regulation of particulate and solution losses from a forest ecosystem. BioScience, *19:*600-610.

Bormann, F. H., G. E. Likens, D. W. Fisher, and R. S. Pierce. 1968. Nutrient loss accelerated by clear-cutting of a forest ecosystem. Science, *159:*882-884.

Braun, E. Lucy. 1947. Development of the deciduous forests of eastern North America. Ecol. Monogr., *17:*211-219.

Braun, E. Lucy. 1950. *Deciduous Forests of Eastern North America.* Blakiston, Philadelphia.

Chaney, R. W. 1940. Tertiary forests and continental history. Bull. Geol. Soc. Amer., *51:*469-488.

Chaney, R. W. 1947. Tertiary centers and migration routes. Ecol. Monogr., *17:* 139-148.

Clements, F. E. 1916. Plant succession: An analysis of the development of vegetation. Carnegie Inst. Washington Publ. *242:*1-512.

Clements, F. E. 1936. Nature and structure of the climax. J. Ecol., *24:*252-284.

Cloud, P. E., Jr. 1968. Atmospheric and hydrospheric evolution on the primitive earth. Science, *160:*729-736.

Danforth, W. F. 1962. Substrate assimilation and heterotrophy. *In* R. R. Lewin [ed.], *Physiology and Biochemistry of Algae,* pp. 99-128. Academic Press, New York.

Dansereau, P. 1957. *Biogeography: An Ecological Perspective.* Ronald, New York.

Dobzhansky, T. 1950. Evolution in the tropics. Amer. Sci., *38:*209-221.

Droop, M. R. 1962. Organic micronutrients. *In* R. A. Lewin [ed.], *Physiology and Biochemistry of Algae,* pp. 141-159. Academic Press, New York.

Ehrlich, P. R., and L. C. Birch. 1967. The "balance of nature" and "population control." Amer. Nat., *101:*97-107.

Ehrlich, P. R., and P. H. Raven. 1964. Butterflies and plants: a study in coevolution. Evolution, *18:*586-608.

Fischer, A. G. 1960. Latitudinal variations in organic diversity. Evolution, *14:* 64-81.

Fogg, G. E., C. Nalewajko, and W. D. Watt. 1965. Extracellular products of phytoplankton photosynthesis. Proc. Roy. Soc. London, Ser. B, *162:*517-534.

Fraenkel, G. S. 1959. The raison d'être of secondary plant substances. Science, *129:*1466-1470.

Gause, G. F. 1934. *The Struggle for Existence.* Reprinted 1964, Hafner, New York.

Gleason, H. A. 1926. The individualistic concept of the plant association. Bull. Torrey Bot. Club, 53:7-26.

Hairston, N. G., F. E. Smith, and L. B. Slobodkin. 1960. Community structure, population control, and competition. Amer. Nat., 94:421-425.

Hardin, G. 1960. The competitive exclusion principle. Science, 131:1292-1297.

Harper, J. L. 1967. A Darwinian approach to plant ecology. J. Ecol., 55:247-270.

Harper, J. L. 1968. The regulation of numbers and mass in plant populations. In R. C. Lewontin [ed.], Population Biology and Evolution, pp. 139-158. Syracuse University Press, Syracuse.

Hartman, R. T. 1960. Algae and metabolites of natural waters. In C. A. Tryon, Jr., and R. T. Hartman [eds.], The ecology of algae, pp. 38-55. Spec. Publ. 2. Pymatuning Lab. Field Biol., University of Pittsburgh.

Hutchinson, G. E. 1959. Homage to Santa Rosalia, or why are there so many kinds of animals? Amer. Nat., 93:145-159.

Hutchinson, G. E. 1961. The paradox of the plankton. Amer. Nat., 95:137-145.

Imbrie, J., and N. Newell [eds.]. 1964. Approaches to Paleoecology. Wiley, New York.

Ladd, H. S. [ed.]. 1957. Treatise on marine ecology and paleoecology. Vol. 2. Paleoecology. Geol. Soc. Amer. Mem., 67:1-1077.

Laporte, L. F. 1968. Ancient Environments. Prentice-Hall, Englewood Cliffs, N. J.

Likens, G. E., F. H. Bormann, N. M. Johnson, D. W. Fisher, and R. S. Pierce. 1970. Effects of forest cutting and herbicide treatment on nutrient budgets in the Hubbard Brook watershed-ecosystem. Ecol. Monogr., 40:23-47.

Lucas, C. E. 1947. The ecological effects of external metabolites. Biol. Rev., Cambridge Phil. Soc., 22:270-295.

MacArthur, R. 1955. Fluctuations in animal populations, and a measure of community stability. Ecology, 36:533-536.

MacArthur, R. H. 1965. Patterns of species diversity. Biol. Rev., Cambridge Phil. Soc., 40:510-533.

Margalef, R. 1963. On certain unifying principles in ecology. Amer. Nat., 97:357-374.

Margalef, R. 1969. Diversity and stability: a practical proposal and a model of interdependence. Brookhaven Symp. Biol., 22:25-37.

Mason, H. L. 1947. Evolution of certain floristic associations in western North America. Ecol. Monogr., 17:201-210.

Muller, C. H. 1940. Plant succession in the Larrea-Flourensia climax. Ecology, 21:206-212.

Muller, C. H. 1952. Plant succession in arctic heath and tundra in northern Scandinavia. Bull. Torrey Bot. Club, 79:296-309.

Odum, E. P. 1959. Fundamentals of Ecology, 2nd ed. Saunders, Philadelphia.

Odum, E. P. 1962. Relationships between structure and function in the ecosystem. Japan. J. Ecol., 12:108-118.

Odum, E. P. 1969. The strategy of ecosystem development. Science, 164:262-270.

Phillips, J. F. V. 1934-5. Succession, development, the climax, and the complex organism: an analysis of concepts. Parts I-III. J. Ecol., 22:554-571, 23:210-246, 488-508.

Ramensky, L. G. 1924. Die Grundgesetzmässigkeiten im Aufbau der Vegetationsdecke. (Russian) Wjestn. opytn. djela Woronesch, 37 p. (Bot. Centbl. N. F., 7:453-455, 1926).

Riley, G. A. 1965. A mathematical model of regional variations in plankton. Limnol. Oceanogr., *10* (Suppl.):R202-R215.

Ryther, J. H. 1963. Geographic variations in productivity. *In* M. N. Hill [ed.], *The Sea,* Vol. 2, pp. 347-380. Interscience, London.

Sanders, H. L. 1968. Marine benthic diversity: a comparative study. Amer. Nat., *102*:243-282.

Sanders, H. L. 1969. Benthic marine diversity and the stability-time hypothesis. Brookhaven Symp. Biol., *22*:71-81.

Shreve, F. 1942. The desert vegetation of North America. Bot. Rev., *8*:195-246.

Simpson, G. G. 1969. The first three billion years of community evolution. Brookhaven Symp. Biol., *22*:162-177.

Slobodkin, L. B. 1962. *Growth and Regulation of Animal Populations.* Holt, Rinehart and Winston, New York.

Slobodkin, L. B., F. E. Smith, and N. G. Hairston. 1967. Regulation in terrestrial ecosystems and the implied balance of nature. Amer. Nat., *101*:109-124.

Solomon, M. E. 1949. The natural control of animal populations. J. Anim. Ecol., *18*:1-32.

Steele, J. H. 1958. Plant production in the northern North Sea. Marine Research, Scientific Home Dept., *1958*(7):1-36.

Strickland, J. D. H. 1965. Production of organic matter in the primary stages of the marine food chain. *In* J. P. Riley and G. Skirrow [eds.], *Chemical Oceanography,* Vol 1, pp. 477-610. Academic Press, New York.

Wallace, B., and A. M. Srb. 1961. *Adaptation.* Prentice-Hall, Englewood Cliffs, N. J.

Wang, Chi-Wu. 1961. The forests of China, with a survey of grassland and desert vegetation. Maria Moors Cabot Foundation, Publ. *5*:1-313. Harvard Univ., Cambridge.

Went, F. W., and N. Stark. 1968. Mycorrhiza. BioScience, *18*:1035-1039.

Whittaker, R. H. 1951. A criticism of the plant association and climatic climax concepts. Northwest Sci., *25*:17-31.

Whittaker, R. H. 1953. A consideration of climax theory: the climax as a population and pattern. Ecol. Monogr., *23*:41-78.

Whittaker, R. H. 1956. Vegetation of the Great Smoky Mountains. Ecol. Monogr., *26*:1-80.

Whittaker, R. H. 1957. Recent evolution of ecological concepts in relation to the eastern forests of North America. Amer. J. Bot., *44*:197-206.

Whittaker, R. H. 1960. Vegetation of the Siskiyou Mountains, Oregon and California. Ecol. Monogr., *30*:279-338.

Whittaker, R. H. 1961a. Vegetation history of the Pacific Coast states and the "central" significance of the Klamath Region. Madrono, *16*:5-23.

Whittaker, R. H. 1961b. Estimation of net primary production of forest and shrub communities. Ecology, *42*:177-180.

Whittaker, R. H. 1965. Dominance and diversity in land plant communities. Science, *147*:250-260.

Whittaker, R. H. 1966. Forest dimensions and production in the Great Smoky Mountains. Ecology, *47*:103-121.

Whittaker, R. H. 1967. Gradient analysis of vegetation. Biol. Rev., Cambridge Phil. Soc., *42*:207-264.

Whittaker, R. H. 1969. Evolution of diversity in plant communities. Brookhaven Symp. Biol., *22*:178-196.

Whittaker, R. H. 1970a. *Communities and Ecosystems.* Macmillan, New York.

Whittaker, R. H. 1970b. The biochemical ecology of higher plants. *In* E. Sond-heimer and J. B. Simeone [eds.], *Chemical Ecology,* pp. 43-70. Academic Press, London.

Whittaker, R. H., and P. P. Feeny. 1971. Allelochemics: chemical interactions between species. Science, *171*:757-770.

Whittaker, R. H., and W. A. Niering. 1965. Vegetation of the Santa Catalina Mountains, Arizona (II). A gradient analysis of the south slope. Ecology, *46*:429-452.

Whittaker, R. H., and W. A. Niering. 1968. Vegetation of the Santa Catalina Mountains, Arizona. IV. Limestone and acid soils. J. Ecol., *56*:523-544.

Whittaker, R. H., and G. M. Woodwell. 1967. Surface area relations of woody plants and forest communities. Amer. J. Bot., *54*:931-939.

Whittaker, R. H., and G. M. Woodwell. 1968. Dimension and production relations of trees and shrubs in the Brookhaven forest, New York. J. Ecol., *56*:1-25.

Whittaker, R. H., and G. M. Woodwell. 1969. Structure, production and diversity of the oak-pine forest at Brookhaven, New York. J. Ecol., *57*:155-174.

Woods, F. W., and R. E. Shanks. 1959. Natural replacement of chestnut by other species in the Great Smoky Mountains National Park. Ecology, *40*: 349-361.

Woodwell, G. M. 1962. Effects of ionizing radiation on terrestrial ecosystems. Science, *138*:572-577.

Woodwell, G. M. 1967. Radiation and the patterns of nature. Science, *156*:461-470.

Woodwell, G. M. 1970. Effects of pollution on the structure and physiology of ecosystems. Science, *168*:429-433.

Woodwell, G. M., and R. H. Whittaker. 1968a. Effects of chronic gamma irradiation on plant communities. Quart. Rev. Biol., *43*:42-55.

Woodwell, G. M., and R. H. Whittaker. 1968b. Primary production in terrestrial ecosystems. Amer. Zool., *8*:19-30.

Discussion

QUESTION: You speak of ecosystems developing toward a steady state. With regard to man's place in and use of ecosystems, can we ever really attain such a steady-state relationship with man continuing to support a life style that takes an ever-increasing amount of energy and materials from the system? And won't this really lead to a retrogression rather than an evolution of many natural ecosystems?

DR. WHITTAKER: It may indeed lead to retrogression—not merely of natural ecosystems but also of what we know as civilization. I think most of you know that it is expected that energy use in the United States should increase by about seven percent a year, and hence should double about every 10 years and expand eightfold by the end of the century. About half that increase should be in fossil fuel combustion. At present levels of fuel combustion, atmospheric pollution is increasing, on a subcontinental and even northern-hemisphere scale, at an accelerating and alarming rate, and with already measurable effects on human health in cities. I do not think we can continue to the end of the century with a three- or four-fold expansion in the sources of pollution problems that are already rapidly outpacing any realistic prospects of control.

I do not like to speak as a prophet of early doom, but let me simply say that this idea of expanding population and expanding industrial expenditure per individual, projected indefinitely on a geometrically steepening trajectory into the future, cannot work. We must then look for a more realistic alternative. I think we must attempt a *strategy* relating man to his environment; a strategy, not piecemeal tactics, because local approaches to these problems are simply overrun by accelerating and increasingly widespread pollution effects. We need if possible a strategy for man's relation to the biosphere, to make possible a long-term occupancy of this world of ours. I think to succeed that strategy must include population stabilization or reduction, a stable, less luxurious but more equitably distributed economic function, effective limitations on environmental degradation by industry, and some consideration of the effects of wealth and commercialism on psychology. We are talking about creating a steady-state system for the occupation of the world by man, and there could hardly be a more profound philosophic revision for our society. We need to change a life style that we in America are still indulging but no longer enjoying. I hope that we may find our way to it.

QUESTION: Population growth isn't really the problem is it? We have fewer people per square mile than many countries, including those of western Europe.

DR. WHITTAKER: Density, average numbers of people per square mile, is just not the point of the population problem for the United States. We can provide housing lots and food for a good many more people than we have. But consider these other, less direct implications of population growth: There is accelerating urbanization, with city size, urban problems, and cost of services all increasing faster than the overall growth rate for the country and the financial resources of cities. Accelerating environmental degradation and toxication are produced by the combination of population growth and increase of industrial wealth. The increase of wealth may seem to have a larger effect on pollution than population growth does, but note that we cannot effectively restrain industrial growth while our economy is straining to provide employment and goods for a growing population. Psychological effects include diminishment of the individual by crowding, congestion, and submergence in the numbers of large organizations and a mass society of powerful technology, a society whose course now seems beyond control by reason based on human welfare and mental health. We experience diminishment also in real wealth by inflation of some costs—notably land and some scarcer resources—while expense of the urban and environmental problems needing solution increases much faster than the economy grows.

The implication of population growth for Americans is not density or food but this whole, malign complex of interrelated environmental, psychological, and economic effects. I suggest the central point is the increasing imbalance between problems and means of decision and solution, the progressive overloading of the society with problems. These problems, some of them intensified by cultural evolution, can hardly fail to increase while our population grows. It is hard, however, to find realistic grounds for thinking that we shall succeed in voluntary population control in this century—it will take a major change in our values.

QUESTION: Would you please comment then on how we might go about implementing a change in our social values?

DR. WHITTAKER: I am sorry I really do not see a way for the present. The values of our society are so strongly dominated by the market place, by television, and by the operation of our communications for the purpose of selling things by creating desires that involve increasing industrial expenditure. Also, parts of the commercial system are more than willing to sell young people clothes and records that represent pseudorebellion against the commercial system, and movies and literature encouraging incompetence and destructiveness toward self and society. Some of the "new people" are not, I am afraid, really in rebellion against commercialism, but are people made psychologically thin and inwardly impoverished by growth in a commercial society, in rebellion against the demands for adult realism, responsibility, and competence judged by others of a high and free civilization. The "counterculture," as some are now calling it, becomes not humanizing progress but the commercially promoted decadence of our society. I do not see how to bring about, in the face of commercial dominance of our culture and the weakness of its victims, a social order valuing both a more frugal way of life and the essential self-disciplines of civilization.

In some ways Western Civilization may have become self-destructive. We Americans have, since the warnings of 1947-1950, indulged a senseless population growth that has left our land increasingly debased and our society saddled with urban problems we really have no idea how to solve—compounded as they are by pollution, change in culture and political temper, and race. The fact that we are releasing into the environment ever-increasing quantities of an ever-increasing variety of toxic materials, with consequences already disturbing if not immediately dangerous, really does imply a general toxication of the biosphere if we keep on. The effects of wealth and technology, intensified by commercial erosion of culture, are producing a reaction much as foretold by Ortega y Gasset (*The Revolt of the Masses*)—people who take the benefits of high civilization for granted as given and their due, and who strike in contempt against the institutions and ideals of personal tolerance and service that make a high civilization work and keep its members employed and fed. These interlinking processes of economic and cultural evolution are not easily changed. But I want to conclude by saying not that we are trapped, but that we will have to change our ways to escape what we are doing to our future. If we are to succeed, we must first seek understanding.

Final Discussion

Led by EUGENE P. ODUM

DR. ODUM: There are several aspects of our comments on ecosystems which I think bear further discussion. First, I would stress the absolute necessity for the wide recognition of the ecosystem concept as an essential approach in ecology, particularly in relation to the problems of man. We simply must look at systems and environments and man more as a unit than we have been doing. If we are to do this, however, we must assess the adequacies and inadequacies of our conception of the ecosystem. We started with the idea of the growth ecosystem and the steady-state ecosystem, or in a cybernetic sense, the transient and the steady-state systems, and Dr. Whittaker has eloquently concluded with exactly the same idea. There is a tremendous challenge in going from man's present growth ecosystem to a steady-state ecosystem, and we can learn very much, as both Dr. Whittaker and Dr. Golley pointed out, by looking at how nature accomplishes the same thing.

Another point which was brought out so often relates to this. The further "up hill" a community or ecosystem is, in terms of its succession or development, the more important are the biological mechanisms for recycling and maintaining stability. This brings out again the sharp differences between terrestrial and marine ecosystems. Ocean ecosystems are dominated by physical forces such as currents and other physical factors. Mature terrestrial systems, on the other hand, are biologically controlled, so the terrestrial system does make a better model for man, because man is the biological control, ultimately, in the human system. So we may contrast the oceans as one extreme system and the great forest as the other, knowing that everything in between has an intermediate relationship. I think that this is a common concept and one that is very useful to think about.

161

A final point, which came out particularly in the last two papers, is that the structure of communities or the structure of the biological parts of communities is not under single control but a result of genetic interactions at the species level. I would hold, however, that the function of the system is under a more central control—that there are control mechanisms of function that operate at the community or ecosystem level. Man is more and more taking over central control of ecosystems, and this has little to do with his genetics.

These are some points which are consistent with contemporary ecosystem theory. What then, are the inadequacies of our theory?

Dr. Golley: One inadequacy is that we really know very little. We have no comprehensive study of all of the components of an ecosystem, although such studies are now beginning as part of the International Biological Program. We do not have a complete inventory of the components of any ecosystem.

Dr. Odum: That's my next topic. I'm talking now about theory, not about getting data and how we do that.

Dr. Golley: All right, but the theory is based upon the facts! The theories of diversity and theories of function are in part influenced by a feeling that we do not really know our system in detail. We have components which we consider "black boxes" which conceal a tremendous amount of variability. Further, ecosystems are not bounded; they are, as Dr. Whittaker and Dr. Likens indicated, continuous, and this gives us some real theoretical problems.

Dr. Odum: But according to the law of integrative control, if you add something to something else you reduce the number of inclusive components. Thus if you add together communities you eventually get the biosphere. Thus by integrating one part into others you simplify in some ways some of the problems of function. For example, it is much easier to measure the CO_2 in the atmosphere and to discuss its interchange with something like the ocean than it is to consider all the details of this interchange.

Dr. Pielou: You wanted me to say some things critical of theory. Well, I'm now going to criticize. I feel that insufficient distinction is made in ecological theorizing between inductive and deductive argument, that people do not say, as I think they should, "I am now going to argue inductively," or "I am now going to argue deductively." If you are going to argue inductively, you must assemble your data, *infer* your conclusion, and say how widely applicable you think it is. If you are going to argue deductively, you must list the assumptions you made and the reasons for which you made them. But so often an ecologist says that his theory is thus and so, and goes on from that point without explaining how and why he arrived at it. I think further detail is needed to explain the underlying thought processes which led an ecologist to propound a theory.

Dr. Whittaker: So large a share of our field is inductive that many of us do not bother to emphasize this most fundamental distinction. Dr. Pielou, in an area of mathematics, was able to use a deductive approach, whereas most of us are dealing with those poorly bounded problems and imperfect relationships in which one largely takes the inductive approach for granted. Successes in applying the deductive approach to community problems are gratifying, but these successes are not extensive.

What is most lacking in the application of ecology to human affairs is, it seems to me, not to discount theory, really good information on transfer mechanisms and magnitudes of pools in the biosphere. We know nothing about the rate at which DDT is removed from the ocean into sediments, and therefore, little about the rate in which DDT and its relatives will accumulate in the ocean. On the carbon dioxide problem in relation to the atmosphere and temperatures one finds oneself estimating here and interpolating there, with no assurance whether warming by the carbon dioxide greenhouse effect or cooling by particle reflection of sunlight will prevail. Thus we find ourselves being "soft scientists"—one trusts the adjective refers to the circumstances of our science and not our heads—on problems on which we would like very much indeed to state hard, realistic predictions. Because we cannot dissuade our society from courses I have suggested are self-destructive, then we cannot persuade our society that our predictions are strong.

Dr. Odum: A great deal of this is involved in our approach to modeling. It has been suggested that if a model is very precise then it is not realistic. And if it's very realistic, then it can't be precise. This is, in a sense, a defense of "soft" science. For instance, universities are attacked by some who hold that they are getting more and more precise about less and less important things. Frequently the theses that graduate students do are more and more precise, but the questions that they are asking are less and less important. Where are the tentative answers on the big questions? I think that we will have to bear in mind that "soft" science may not be as precise as "hard" science, but it is just as vitally important, since the decisions which must be made have to be based on a realistic model. That means that we are going to have to decide not in a precise way—but in a realistic or tentative way— how the world must be managed. We are going to have to determine how industry should interact with political science and economics, and that's pretty soft stuff. So let's not get discouraged because ecology is not a "hard" science; it is an exceedingly important science. With that in mind, it is apparent that we need not so much more theory but better application; therefore, we do not need to train a lot of people exclusively in ecological theory, but we do need more specialists who understand the whole and can obtain the data that we can put together to form a realistic model. There is a terrible inadequacy in our data, particularly in basic inventories of ecosystem components and rates of exchange between components.

DR. LIKENS: I, for one, feel that we have an adequate number of basic hypotheses and theories at the ecosystem level to attack complex ecological problems. More importantly, we now are learning a great deal about how to obtain quantitative information about ecosystems, and where critical answers are necessary to test the hypotheses. Until we know what the relationships are and at least order-of-magnitude values for these functional interactions between the structural components of the ecosystem, important questions won't emerge; and unless we ask the right questions our hypotheses won't improve. One of the vital ecosystem components, in this regard, is the so-called decomposers. The role of these microconsumers in the functioning of ecosystems is a critical area about which we know very little, relative to other components. We've known for a very long time that things decay, and we've known for a very long time that there are bacteria and fungi in various environments, but only very recently have we gotten even order-of-magnitude values for the importance of these niches to total ecosystem function. We now find that they are extremely important and in many cases are the vital links in both energy flow and nutrient cycling in ecosystems.

DR. GOLLEY: Let me add another point. We've organized our models of ecosystems partly on physical grounds such as the physical effect of eating and transfers of food and energy and materials from one compartment to another. There are also, as Bob Whittaker pointed out, the chemical influences. These are more easily investigated in aquatic systems and, of course, have been studied to some extent. I would suspect that there is a web of chemical relationships equally as complicated as the web of trophic relationships, and these influences strongly affect the behavior of species populations. I predict that in the next 10 years we will see this particular area of ecology developing as actively as the area of trophic relationships did in the past 10 years.

DR. ODUM: Another area which should be mentioned is the impact of technology on ecology. We tend to think of technology as creating only ecological ills, but some recent technological achievements raise our potential for conducting really intensive ecological analyses of ecosystems. Remote sensing technology, for example, gives us the ability to get up above the biosphere or ecosystems, to look at a whole forest or a whole lake or even measure the temperature of the ocean simply by sensing with various kinds of photography. We now have sensors capable of coding many kinds of information emanated as waves or rays. This way we will be able to record carbon dioxide concentrations over large areas, for example. Another great breakthrough is in continuous monitoring techniques. For example, with carbon sensors all over the world continuously monitoring concentrations, we would know if it started to go up or down. As Dr. Whittaker pointed out, we are now guessing about a lot of these things. A third area of technological contribution is in tracer techniques. For instance, in determining the ecological relationships between organisms, we can use tracers to see whether they are exchanging energy, food, or materials. I know how surprised we were when we tagged a couple of species of plants with radioactive phosphorous. There were some 150 kinds of insects which could be collected from these plants, but only about 15 of them picked up any appreciable radioactivity. Thus, very few of the insect species were very closely interacting with the two species of plants; the others were either just using them as a perch, or feeding on some other plants, or had some other nonfeeding relationship with the plants. Finally, system analysis techniques, which were originally developed by engineers, hold great potential for modeling, testing, and simulating ecological theories.

Dr. RILEY: The application of new technology to oceanography has made it possible to work on problems which we knew existed but simply did not know how to deal with by older methods. For example, it has become increasingly evident in recent years in the study of the ocean that we should be more involved with the trace quantities of organic substances that occur. These are part of the primary production; they are used by bacteria and ultraplankton and are liberated in the excretion of animals. There are thus several ways that such trace organic substances can be taken up by the biological system, and at least some of this organic matter can be converted into particulate form and be transferred to higher trophic levels of the food chains. We have not had the technology to deal with this problem until a very few years ago. The development of the gas chromatograph and various other things has opened up a whole field which is really a very much larger segment of biological oceanography than we realized until very recently.

Dr. ODUM: Another thing I'd like to mention is the tremendously exciting international efforts being made to collect data and to put the ecosystem theory into effect. One of these is the International Biological Program, which is now operating in nearly every country in the world. This program is based on the concept of studying whole ecosystems, and in this country we are basing it on the concept of the watershed, as exemplified by the studies of Dr. Likens and his colleagues on Hubbard Brook. The IBP is now doing what oceanographers have already well started; oceanography is by nature an international science and there is a great deal of international cooperation. In addition to this there are the physical monitorings of ecosystems which are now well along, including such programs as the International Hydrological Decade and the International Geophysical Year. So there are underway massive attempts by mankind to try to obtain the necessary information to apply ecosystem theory.

The final point that I'd like to raise is the question of how we develop an application of ecosystem theory to man—how do we develop an applied human ecology? We've had a human ecology in the past only in terms of describing man in relation to his environment, but human ecologists have not considered the sort of complex oscillating steady-state balance between man and nature which is needed. We need to manage our own population as well as those of the organisms on which we depend or exploit. Universities traditionally have a department of wildlife management, a department of forest management, and so on, but where are the departments of human management? How do we find a way to manage our population as part of the ecosystem; what kinds of procedures are necessary or possible? Does anyone care to tackle this very difficult subject? Bob?

DR. WHITTAKER: I wasn't volunteering, you notice. We have really needed for some time a new science of human ecology comprising at least study of human populations, the interactions of industrial man and the biosphere, and economic alternatives including steady-state function. For the present many economists abhor the idea of a stable population and economy; they even think the idea as "unrealistic" as we may consider continued growth to be. But I think we should try to bring about a field of study that would relate these things to one another—population and biosphere, resources and economic function, cultural morale and political means—in the hope that from such study there might emerge an integrated strategy for the human future.

DR. GOLLEY: It seems to me that there are some basic philosophical problems here. We have on the one hand a philosophical position which postulates the evolution of man, in which man began at some point and is continually progressing into the future. Another philosophical position is that we operate in cycles, that we go through a never-ending cyclic sequence of steps repeating again and again the same patterns of our relations with each other and with our environment.

Then there is a third rather existential position where, as Slobodkin has said, we are playing an existential game in which the object is to keep playing the game. Depending upon one's philosophical position, one's attitude toward human ecology and environmental problems may be very different. I would argue that the latter position makes biosphere sense and ecosystem sense. The other two lead us into traps, either postulating greater control, which I would argue is suicidal in its largest conception, or postulating cycles, which is more or less fatalistic.

DR. ODUM: Again, I think that, with regard to pollution and so on, that there are two basic strategies. One is to correct the mistakes or treat the symptoms that we already have. The other one, more important in my opinion, is the prevention of further deterioration of a quality environment. I think the thing that impresses me is that a great many people, not just scientists, but writers, economists, and even some politicians, are beginning to say the same thing. For instance, Lewis Mumford in his many writings as a social critic makes the statement that the greatest challenge of our age is how quality can control quantity. This suggests one basis for a strategy, then—put the emphasis on the quality of the individual or the quality of the product or the quality of the environment and we may then have a strategy similar to that of nature. I'm a little discouraged sometimes by the way the government goes at this, because all of the action and reaction is directed towards penalizing somebody for polluting or attacking industry for polluting in the past. The real problem and need is to see that we do not go further down the pollution road in the immediate future. If waste management could become the first priority in all new development (and not the last consideration as in present planning), then we could emphasize quality and reasonable standards from the outset, not in retrospect.

DR. LIKENS: It seems that we're all saying we need a system based on real or true economics, and that what we are doing now is basing our decisions, and in fact the system, on faulty economics. We currently have a ledger system which is entirely based upon man, which is egocentric, and what we really should do is construct a balance sheet or ledger based upon all costs associated with the functioning of the biosphere, or individually upon entire ecosystems. When you do that, you end up with an honest balance sheet, in which you include all the ecological costs as well as the monetary cost of the product. Then more realistic ecological decisions can be made on whether some manipulation or endeavor is really economically feasible or worth it.

DR. GOLLEY: When you say it's worth it, there is still a value judgment to be made. I agree with you that accurate economic accounting is essential, but we still must make decisions on the basis of our values.

Dr. Whittaker: Dr. Likens has observed the necessity of a more honest balancing of the costs, and benefits of our tactical moves in relation to environment. I agree heartily and would reinforce his comment with this warning: We cannot solve our problems by a balance sheet based only on economic measurements. We cannot really solve the problems by balancing the costs in terms of indirect and unexpected environmental degradation, and still more subtle and indirect costs to human psychology, against direct economic benefits. There are reasons we cannot do this successfully. The first is the fact that the costs and benefits are incommensurable with one another. A second reason is the unpredictableness and the unforeseen acceleration of the costs, and the fact that the costs bear later on different people than the benefits. A third reason is that growth of population and wealth increases the economic pressures toward environmental degradation without proportionately increasing the measurable costs. A ledger that supports recreational or agricultural use of land at one time becomes, only a few years later with population growth, a ledger in which short-term profits of development are decisive. We know that persistent chlorinated hydrocarbons are destructive to man's long-term interests, but we cannot expect the leadership of an increasingly hungry poor country to decide against the direct benefits to themselves, for the sake of later costs to the world at large. Thus we are, again, boxed in by population growth.

This relates to what Dr. Golley was calling an existential game. We are indeed playing a kind of game. The popular name for it is double or nothing. One does not win at double or nothing, beyond a certain point. We might be there. I think then we must look to more than a more honest balancing of costs and benefits, toward change in strategy. We must bring ourselves to a different game, and while I could not overstate the difficulty, I find encouraging the increasing awareness in this country that the game we're playing can really lose unless its rules are changed.

Appendix

Thirty-first Annual Biology Colloquium

Theme: Ecosystem Structure and Function

Dates: April 24-25, 1970

Place: Oregon State University, Corvallis, Oregon

Special Committee for the 1970 Biology Colloquium: John A. Wiens *co-chairman,* C. David McIntire, *co-chairman,* William K. Ferrell, William P. Nagel, W. Scott Overton, Lawrence A. Small

Standing Committee for the Biology Colloquia: J. Ralph Shay, *chairman,* Paul O. Ritcher, Paul R. Elliker, Ernst J. Dornfeld, Henry P. Hansen, Hugh F. Jeffrey, J. Kenneth Munford, John M. Ward.

Colloquium Speakers, 1970:

Eugene P. Odum, Alumni Foundation Distinguished Professor and Director, Institute of Ecology, University of Georgia, *leader*

Gene E. Likens, Associate Professor, Section of Ecology and Systematics, Cornell University

Frank B. Golley, Executive Director, Institute of Ecology, University of Georgia

Gordon A. Riley, Director, Institute of Oceanography, Dalhousie University

E. C. Pielou, Professor, Department of Biology, Queen's University

Robert H. Whittaker, Professor, Section of Ecology and Systematics, Cornell University

Sponsorship:

Agricultural Experiment Station, Oregon State University

Environmental Health Sciences Center, Oregon State University

Graduate Research Council, Oregon State University

Oregon State University Chapter, Ecological Society of America

Phi Kappa Phi

School of Forestry, Oregon State University

School of Science, Oregon State University

Sigma Xi

Index